Wissenschaftliche Reihe
Fahrzeugtechnik Universität Stuttgart

Reihe herausgegeben von
M. Bargende, Stuttgart, Deutschland
H.-C. Reuss, Stuttgart, Deutschland
J. Wiedemann, Stuttgart, Deutschland

Das Institut für Verbrennungsmotoren und Kraftfahrwesen (IVK) an der Universität Stuttgart erforscht, entwickelt, appliziert und erprobt, in enger Zusammenarbeit mit der Industrie, Elemente bzw. Technologien aus dem Bereich moderner Fahrzeugkonzepte. Das Institut gliedert sich in die drei Bereiche Kraftfahrwesen, Fahrzeugantriebe und Kraftfahrzeug-Mechatronik. Aufgabe dieser Bereiche ist die Ausarbeitung des Themengebietes im Prüfstandsbetrieb, in Theorie und Simulation. Schwerpunkte des Kraftfahrwesens sind hierbei die Aerodynamik, Akustik (NVH), Fahrdynamik und Fahrermodellierung, Leichtbau, Sicherheit, Kraftübertragung sowie Energie und Thermomanagement – auch in Verbindung mit hybriden und batterieelektrischen Fahrzeugkonzepten.

Der Bereich Fahrzeugantriebe widmet sich den Themen Brennverfahrensentwicklung einschließlich Regelungs- und Steuerungskonzeptionen bei zugleich minimierten Emissionen, komplexe Abgasnachbehandlung, Aufladesysteme und -strategien, Hybridsysteme und Betriebsstrategien sowie mechanisch-akustischen Fragestellungen.

Themen der Kraftfahrzeug-Mechatronik sind die Antriebsstrangregelung/Hybride, Elektromobilität, Bordnetz und Energiemanagement, Funktions- und Softwareentwicklung sowie Test und Diagnose.

Die Erfüllung dieser Aufgaben wird prüfstandsseitig neben vielem anderen unterstützt durch 19 Motorenprüfstände, zwei Rollenprüfstände, einen 1:1-Fahrsimulator, einen Antriebsstrangprüfstand, einen Thermowindkanal sowie einen 1:1-Aeroakustikwindkanal.

Die wissenschaftliche Reihe „Fahrzeugtechnik Universität Stuttgart" präsentiert über die am Institut entstandenen Promotionen die hervorragenden Arbeitsergebnisse der Forschungstätigkeiten am IVK.

Reihe herausgegeben von

Prof. Dr.-Ing. Michael Bargende
Lehrstuhl Fahrzeugantriebe,
Institut für Verbrennungsmotoren und
Kraftfahrwesen, Universität Stuttgart
Stuttgart, Deutschland

Prof. Dr.-Ing. Jochen Wiedemann
Lehrstuhl Kraftfahrwesen,
Institut für Verbrennungsmotoren und
Kraftfahrwesen, Universität Stuttgart
Stuttgart, Deutschland

Prof. Dr.-Ing. Hans-Christian Reuss
Lehrstuhl Kraftfahrzeugmechatronik,
Institut für Verbrennungsmotoren und
Kraftfahrwesen, Universität Stuttgart
Stuttgart, Deutschland

More information about this series at http://www.springer.com/series/13535

Oliver Fischer

Investigation
of Correction Methods
for Interference Effects in
Open-Jet Wind Tunnels

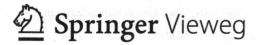 Springer Vieweg

Oliver Fischer
Stuttgart, Germany

Dissertation University of Stuttgart, 2017

D93

Wissenschaftliche Reihe Fahrzeugtechnik Universität Stuttgart
ISBN 978-3-658-21378-7   ISBN 978-3-658-21379-4   (eBook)
https://doi.org/10.1007/978-3-658-21379-4

Library of Congress Control Number: 2018936517

Springer Vieweg

This Springer Vieweg imprint is published by the registered company Springer Fachmedien Wiesbaden
GmbH part of Springer Nature
The registered company address is: Abraham-Lincoln-Str. 46, 65189 Wiesbaden, Germany

Preface

This Ph.D. thesis contains the results of research undertaken at the Institute for Internal Combustion Engines and Automotive Engineering at the University of Stuttgart. This research project was realized with a sponsorship of Exa Corporation.

Certainly, I never would have finished my dissertation without the help and support of others. These years have been a long and winding road, with both ups and downs. Fortunately, I was not alone on this trip, but accompanied by people, who were always willing to coach, support, help, and motivate me. For this, I would like to kindly thank them.

I am especially indebted to Prof. Dr. Jochen Wiedemann. I offer my sincerest gratitude for his trust in me when I first came to his institute being a physicist and willing to do aerodynamics. He constantly supported and stimulated my research, always showing great interest in the progress of my work.

Many thanks also to Prof. Dr. Cameron Tropea for the time spent on evaluating my thesis, as well as for the useful comments and suggestions.

I am grateful to the sponsor of this research project at Exa Corporation, Steve Remondi, who consistently and actively supported the project. And, many thanks to all the people at Exa, who graciously contributed, especially Brad Duncan.

Many thanks also to Nils Widdecke (department head of "Vehicle Aerodynamics and Thermal Management") for his comments, suggestions, and continuous support. Thanks for always taking the time to discuss scientific problems with me.

Furthermore, I extend my gratitude to Dr. Timo Kuthada (department head of "Interdisciplinary Projects and High Performance Computing") for his guidance, committed support, and supervision through the years. Thanks for the countless useful discussions, comments, and suggestions.

A heartfelt thanks to my fellow researchers and friends at the department. I very much enjoyed our daily exchange of ideas and thoughts, as well as just the fun moments we shared. Many thanks also go to all colleagues and students at IVK and FKFS, especially the wind tunnel crews.

I wish to thank all the other people who worked with me over the years on these topics, in particular, Dr. Edzard Mercker and the coauthors of the published papers.

A special thanks to my mother and father for their tireless encouragement, motivation, and support during the years.

Finally, I wish to particularly thank my wife, Christine. She deserves a special word of appreciation for her continuous moral support, her patience, and her love.

Table of Contents

Preface.. V

List of Figures XI

List of Tables XIX

Formula Symbols XXI

List of Abbreviations XXIII

Zusammenfassung.. XXV

Abstract XXVII

1   Introduction 1

   1.1   Aerodynamic Vehicle Development – Experiment and
Simulation.. 1

   1.2   Motivation and Problem 2

2   Aerodynamic Development Tools 5

   2.1   Wind Tunnel Technology 5

      2.1.1   Wind Tunnel Design 6

         2.1.1.1   Wind Tunnel Types 6

         2.1.1.2   Wind Tunnel Test Sections 9

         2.1.1.3   Specific wind tunnel configurations 10

      2.1.2   Wind Tunnel Calibration 11

   2.2   Computational Fluid Dynamics - Numerical Scheme.................. 14

      2.2.1   The Lattice-Boltzmann Method 15

      2.2.2   Fluid Turbulence Model and Wall Model..................... 16

3   Wind Tunnel Interference Effects 19

    3.1   Blockage Effects in Open-Jet Wind Tunnels 20

        3.1.1   Jet Expansion Effect 21

        3.1.2   Nozzle blockage effect 22

        3.1.3   Jet Deflection Effect 23

        3.1.4   Collector Effect 23

    3.2   Effect of Static Pressure Distribution (Horizontal Buoyancy) 23

**4   Correction Method for Interference Effects in Open-Jet
Wind Tunnels 25**

    4.1   Historical Development 26

        4.1.1   Open-Jet Blockage Effects 27

        4.1.2   Horizontal Buoyancy 28

        4.1.3   Mercker-Wiedemann Method 28

    4.2   Dynamic Pressure Correction 31

        4.2.1   Solid Blockage 33

        4.2.2   Collector Effect 34

        4.2.3   Nozzle Effect 36

        4.2.4   Application Points for the Interference Velocities 38

    4.3   Horizontal Buoyancy Correction 39

    4.4   Open-jet correction method 40

5   Computational Fluid Dynamics Investigations 41

    5.1   Basic CFD Investigation of Horizontal Buoyancy 41

    5.2   Validation Measurements in IVK-MWK 45

    5.3   Simulation of IVK-MWK (DIVK) 47

        5.3.1   Digital Wind Tunnel Model (DIVK) 47

        5.3.2   Boundary conditions 53

5.3.3 Boundary Layer.. 55

5.3.4 Shear Layer 56

5.3.5 Static Pressure Distribution............................. 59

5.4 CFD Simulation of Different Vehicle Models 61

5.4.1 SAE Squareback 62

5.4.1.1 Centerline Pressure Distribution..................... 62

5.4.1.2 Total Pressure in Near Field Wake................ 63

5.4.1.3 Forces 65

5.4.2 Detailed Notchback (Scale 1:5) 66

5.4.2.1 Centerline Pressure Distribution................... 67

5.4.2.2 Flow Field... 68

5.4.2.3 Forces 72

5.4.3 Detailed notchback (scale 1:4)...................... 72

5.4.3.1 Flow Field... 73

5.4.3.2 Forces 75

5.4.4 Detailed SUV model (scale 1:4) 77

5.4.4.1 Flow Field... 78

5.4.4.2 Forces 81

5.5 Conclusions.. 85

6 Application and Investigation of the Correction Method.. 87

6.1 Application to Wind Tunnel Measurements 87

6.1.1 Coupe Vehicle in IVK-FWK 88

6.1.2 Van in IVK-FWK — High-Blockage Setup 90

6.1.3 Detailed Notchback (Scale 1:4) in IVK-MWK — Inhomogeneous Static Pressure Distribution 92

6.1.4 Detailed Notchback (Scale 1:5) in IVK-MWK............. 98

6.1.5 Detailed Notchback (Scale 1:4) in IVK-MWK............. 99

        6.1.6   Detailed SUV Model (Scale 1:4) in IVK-MWK........... 100

    6.2 Conclusions... 102

7   **Comparison of Results** **105**

    7.1 Comparison of Computational Fluid Dynamics and Correction
         Results... 105
    7.2 Conclusions... 107
    7.3 Outlook 107

8   **Bibliography**.. **109**

A   **Appendix** **117**

    A.1 Vehicles and Vehicle Models 117
        A.1.1   SAE Squareback Model.................................. 117
        A.1.2   Notchback Model (Scale 1:5)........................... 119
        A.1.3   Detailed Notchback Model (scale 1:4).............. 120
        A.1.4   Detailed SUV Model....................................... 121
        A.1.5   Coupe (full-scale)... 122
        A.1.6   Van (full-scale) 122
    A.2 Wind tunnels.. 123
        A.2.1   IVK Model-Scale Wind Tunnel (IVK-MWK)............... 123
        A.2.2   IVK Full-Scale Aero-Acoustic Wind Tunnel (IVK-
                FWK)   126
    A.3 Application of the Modified Mercker-Wiedemann Correction
         Method 128
        A.3.1   Step 1: Calculation of source point position xS 128
        A.3.2   Step 2: Calculation of Sensitivity Length 131
    A.4 Calculation of Mass Flow for Boundary Conditions in CFD..... 136
    A.5 Flow Field around Notchback Model (scale 1:5)...................... 138

List of Figures

Figure 2.1: Open circuit wind tunnel with an open-jet test section ("Eiffel") [19] 6

Figure 2.2: Closed circuit wind tunnel with an open-jet test section ("Göttingen") [19] 7

Figure 2.3: Wind tunnel test section types (open-jet, closed-walls) [1].......9

Figure 2.4: Determination of dynamic pressure in the wind tunnel test section with nozzle method (left) and plenum method (right)..12

Figure 3.1: Blockage effects in open-jet wind [1].....................................21

Figure 3.2: Undisturbed flow with ambient static pressure in large distance from the model...22

Figure 3.3: Horizontal buoyancy in open-jet wind tunnels [1] 24

Figure 5.1: Simple computational fluid dynamics simulation to investigate horizontal buoyancy (closed-wall wind tunnel and test body)..42

Figure 5.2: Bluff (left) and streamlined (right) test bodies 43

Figure 5.3: Static pressure distributions for the basic computational fluid dynamics investigation of horizontal buoyancy.............44

Figure 5.4: Measured and corrected drag coefficients for the two test bodies simulated in different static pressure distributions.......44

Figure 5.5: Measured static pressure distribution ($z = 50$ mm), rear center belt roller gaps at dotted line...46

Figure 5.6: Measured static pressure distribution ($z = 150$ mm), rear center belt roller gaps at dotted line...46

Figure 5.7: Measured static pressure distribution ($z = 450$ mm), rear center belt roller gaps at dotted line...46

Figure 5.8: Digital model of IVK model-scale wind tunnel with nozzle contraction (left) and diffuser (right).......................................48

Figure 5.9: Influence of different turbulence generator configurations
in IVK-MWK 49

Figure 5.10: Delta-wings (simulation model [left], real geometry
[right]) at nozzle exit plane 49

Figure 5.11: Static pressure distribution using wing boxes as turbulence
generators in DIVK 50

Figure 5.12: Simple cubes were tested to be used as turbulence
generators in DIVK 51

Figure 5.13: Additional ventilation inlets in DIVK (white) 52

Figure 5.14: Computational mesh of test section without vehicle (left
y = 0, right z = 0.45 m): high resolution for shear layer
development and boundary layer 52

Figure 5.15: Computational mesh around delta-wings at nozzle exit 53

Figure 5.16: Comparison of boundary layer in test section simulation
and measurement (configuration 1) 56

Figure 5.17: Comparison of boundary layer in test section simulation
and measurement (configuration 1) 56

Figure 5.18: Simulation of shear layer development along test section,
with detailed delta-wings (configuration 1), displayed
planes at x = ±0.5 m 57

Figure 5.19: Flow structure in the empty test section (configuration 1),
measurement (left) and simulation (right), x = 0.5 m (near
front of model), black line represents nozzle size 58

Figure 5.20: Flow structure in the empty test section (config 1),
measurement (left) and simulation (right), x = -0.5 m (near
rear end of model), black line represents nozzle size 58

Figure 5.21: IVK-MWK collector: left simulation (with/without
stagnation bodies mounted), right experiment with
stagnation bodies mounted 60

Figure 5.22: Comparison of static pressure distributions generated with
stagnation bodies 60

Figure 5.23: Centerline pressure distribution on SAE squareback model
with standard static pressure distribution (configuration 1) 62

Figure 5.24: Centerline pressure distribution on SAE squareback model with standard static pressure distribution (configuration 3) 63

Figure 5.25: SAE squareback model: Total pressure 50 mm behind model base (configuration 1), measurement (left) and simulation [VPP] (right) 64

Figure 5.26: SAE squareback model: Total pressure behind model base (y = 0 mm) (configuration 1), measurement (left) and simulation [VPP] (right) 64

Figure 5.27: SAE squareback model: Integral values of drag and lift from simulation and experiment for both configurations (configuration 2 not available for this model) 66

Figure 5.28: Detailed notchback (scale 1:5): Surface pressure measurements between experiment (configuration 3), DIVK (configuration 3) and DWT, at the centerline (y = 0 mm), top side of the model 67

Figure 5.29: Detailed notchback (scale 1:5): Surface pressure measurements between experiment (configuration 3), DIVK (configuration 3) and DWT, at y = 60 mm, top side of the model 68

Figure 5.30: Notchback (scale 1:5), $C_{p, stat}$ at X = 500 mm plane: DIVK config 1 (top left); Exp config 1 (top right); DIVK config 3 (center left); Exp config 3 (center right); DWT (bottom left), no measurement data in checkered area 70

Figure 5.31: Notchback (scale 1:5), $C_{p, tot}$ at X = 500 mm plane: DIVK config 1 (top left); Exp config 1 (top right); DIVK config 3 (center left); Exp config 3 (center right); DWT (bottom left), no measurement data in checkered area 71

Figure 5.32: Detailed notchback (scale 1:5)—Integral values of drag and lift from simulation and experiment for both configurations (static pressure distribution) 72

Figure 5.33: Notchback (scale 1:4), Cp-static at Y = 0 mm plane in the
near field wake: top row: DIVK config 1 (left); Exp config
1 (right); second row: DIVK config 2 (left); Exp config 2
(right); third row: DIVK config 3 (left); Exp config 3
(right); DWT (bottom row left), no measurement data in
checkered areas 74

Figure 5.34: Detailed notchback (scale 1:4): Drag and lift values for
Experiment, DIVK (configuration 1), and DWT 76

Figure 5.35: Detailed notchback (scale 1:4): Drag and lift values for
Experiment, DIVK (configuration 2), and DWT 76

Figure 5.36: Detailed notchback (scale 1:4): Drag and lift values for
Experiment, DIVK (configuration 3), and DWT 77

Figure 5.37: SUV-RH1 (scale 1:4), Cp-static at Y = 0 mm plane in the
near field wake (no measurement data in striped areas):
top row: DIVK config 1 (left); Exp config 1 (right);
second row: DIVK config 2 (left); Exp config 2 (right);
third row: DIVK config 3 (left); Exp config 3 (right);
bottom row: DWT (left) 79

Figure 5.38: SUV-RH2 (scale 1:4), Cp-static at Y = 0 mm plane in the
near field wake (no measurement data in striped areas):
top row: DIVK config 1 (left); Exp config 1 (right);
second row: DIVK config 2 (left); Exp config 2 (right);
third row: DIVK config 3 (left); Exp config 3 (right);
bottom row: DWT (left) 80

Figure 5.39: Detailed SUV-RH1 (scale 1:4): Drag and lift values for
Experiment, DIVK (configuration 1), and DWT 82

Figure 5.40: Detailed SUV-RH1 (scale 1:4): Drag and lift values for
Experiment, DIVK (configuration 2), and DWT 82

Figure 5.41: Detailed SUV-RH1 (scale 1:4): Drag and lift values for
Experiment, DIVK (configuration 3), and DWT 83

Figure 5.42: Detailed SUV-RH2 (scale 1:4): Drag and lift values for
Experiment, DIVK (configuration 1), and DWT 84

Figure 5.43: Detailed SUV-RH2 (scale 1:4): Drag and lift values for
Experiment, DIVK (configuration 2), and DWT 84

Figure 5.44: Detailed SUV-RH2 (scale 1:4): Drag and lift values for experiment, DIVK (configuration 3), and DWT 85

Figure 6.1: Static pressure distributions in IVK-FWK generated by changing the collector flap angle (without ground simulation).. 89

Figure 6.2: Correction results of coupe measured in IVK-FWK (with and without empirical near field wake constant $f_{nfw} = 0.41$).... 90

Figure 6.3: Van in the test section of IVK-FWK 91

Figure 6.4: Correction results of the van measured in IVK-FWK (with and without the empirical near field wake constant $f_{nfw} = 0.41$).. 91

Figure 6.5: SUV model at several x-positions inside the collector to generate different static pressure distributions in the test section.. 93

Figure 6.6: Static pressure distributions generated with SUV model at several x-positions inside the collector 94

Figure 6.7: Static pressure measurements (x = 700 mm, z = 150 mm); small picture: measurement position (white arrow) relative to vehicles 95

Figure 6.8: Measurement situations in the test section (force measurement on balance on notchback model; SUV model used as stagnation body inside collector) 96

Figure 6.9: Correction results of the notchback model (scale 1:4) in IVK-MWK, in an inhomogenous static pressure distribution (with and without the empirical near field wake constant $f_{nfw} = 0.41$)... 97

Figure 6.10: Correction results of detailed notchback (scale 1:5) measured in IVK-MWK (with and without the empirical near field wake constant $f_{nfw} = 0.41$) 99

Figure 6.11: Correction results of detailed notchback model (scale 1/4) measured in IVK-MWK (with and without the empirical near field wake constant $f_{nfw} = 0.41$) 100

Figure 6.12: Correction results of detailed SUV model (scale 1:4) measured in IVK-MWK—in ride height 1 (with and without the empirical near field wake constant f_{nfw} = 0.41) .. 101

Figure 6.13: Correction results of detailed SUV model (scale 1:4) measured in IVK-MWK—in ride height 2 (with and without the empirical near field wake constant f_{nfw} = 0.41) .. 102

Figure A.1: Picture of SAE squareback model inside IVK-MWK 117

Figure A.2: SAE reference model (dimensions given for full-scale model) [68] 118

Figure A.3: Notchback model (scale 1:5) 119

Figure A.4: Detailed notchback model (scale 1:4)...................................... 120

Figure A.5: Detailed SUV model (scale 1:4) 121

Figure A.6: Renault Master I 122

Figure A.7: IVK model-scale wind tunnel [56] 124

Figure A.8: Nozzle of IVK model-scale wind tunnel 124

Figure A.9: Collector of IVK model-scale wind tunnel............................ 125

Figure A.10: Coordinate system in IVK-MWK... 125

Figure A.11: IVK-FWK [56] 127

Figure A.12: Coordinate system in IVK-FWK 127

Figure A.13: Illustration of the wind tunnel test section with a model present... 134

Figure A.14: Ratio of dynamic pressures q in the empty wind tunnel test section.. 137

Figure A.15: Comparison of the surface pressure measurements between the Exp config 3, DIVK config 3, and DWT, y = 0 mm on the underbody of the model............................. 138

Figure A.16: Comparison of the surface pressure measurements between the Exp config 3, DIVK config 3, and DWT, y = 60 mm on the underbody of the model........................... 139

Figure A.17: Cp-static at Y = 255 mm plane: top row: DIVK config 1 (left); Exp config 1 (right); center row: DIVK config 3 (left); Exp config 3 (right); bottom row: DWT (left)............. 140

Figure A.18: Cp-static at Z = 30 mm plane: top row: DIVK config 1 (left); Exp config 1 (right); center row: DIVK config 3 (left); Exp config 3 (right); bottom row: DWT (left)............. 141

Figure A.19: Cp-static in the wake (Y = 0 plane): Exp config 3 (top left); DIVK config 3 (top right); Digital wind tunnel DWT (bottom left)... 142

Figure A.20: Cp-total in the wake (positive y): Exp config 3 (top left); Wind tunnel todel DIVK config 3 (top right); Digital wind tunnel DWT (bottom left)... 143

Figure A.21: Cp-total in the backlight area (iso-surfaces): Exp config 3 (top left); Wind tunnel todel DIVK config 3 (top right); Digital wind tunnel DWT (bottom left)................................. 144

List of Tables

Table 2.1: Possible combinations of wind tunnel types and test section types 10

Table 5.1: Boundary conditions in DIVK 54

Table 6.1: Correction results of notchback model with two different ways of treating the inhomogeneous static pressure distributions in the correction procedure 97

Table 7.1: Comparison of resulting drag values from DWT simulations and the correction method 106

Table A.1: Technical specifications of SAE squareback model 117

Table A.2: Technical specifications of notchback model (scale 1:5) 119

Table A.3: Technical specifications of detailed notchback model (scale 1:4) 120

Table A.4: Technical specifications of detailed SUV model (scale 1:4).. 121

Table A.5: Technical specifications of MB CLK 122

Table A.6: Technical specifications of Renault Master........................... 123

Table A.7: Technical specifications of IVK-MWK................................. 123

Table A.8: Detailed information on stagnation body positions inside the collector 126

Table A.9: Technical specifications of IVK-FWK 126

Table A.10: Correction method: Overview of calculated x_S 129

Formula Symbols

A	Surface area of test object	[m²]
A_C	Collector cross-section	[m²]
A_M	Model area frontal	[m²]
A_N	Nozzle cross-section	[m²]
CD_{meas}	Measured drag coefficient	-
ΔCD_{HB}	Correction term for horizontal buoyancy	-
$cp_{stat}(x)$	Static pressure distribution at x-Position	-
ε_i	Blockage correction factor	-
ε_S	Solid-blockage correction factor	-
ε_C	Collector effect correction factor	-
$\varepsilon_{N/P}$	Nozzle blockage effect correction factor (N: nozzle method, P: plenum method)	-
ε_{QN}	Nozzle blockage effect correction factor	-
ε_W	Wake blockage effect correction factor	-
f_{nfw}	Near field wake constant	-
f_P	Plenum Factor (plenum method: $f_P = 1$, nozzle method: $f_P = 0$)	-
G	Glauert Factop	-
k	Nozzle factor	-
k_N	Nozzle factor for the nozzle method	-
k_P	Nozzle factor for the plenum method	-
l_M	Model length	[m]
l_{TS}	Wind tunnel test section length	[m]
p_{tot}	Total pressure	[Pa]

p_{stat}	Static pressure	[Pa]
p_{SC}	Pressure in wind tunnel settling chamber	[Pa]
p_N	Pressure in the nozzle contraction close to nozzle exit plane	[Pa]
p_P	Pressure in the wind tunnel plenum	[Pa]
q	Dynamic pressure	[Pa]
r_C	Equivalent radius of the duplex collector area	[m]
r_N	Equivalent radius of the duplex nozzle area	[m]
t	Thickness vehicle model	-
τ	Tunnel shape factor	-
τ_r	Relaxation time parameter	-
U_∞	Measured velocity	[m/s]
U_{corr}	Corrected velocity	[m/s]
Δu_i	Velocity increments for different blockage effects	[m/s]
v	Velocity	[m/s]
V	Volume of test object	[m³]
V_{eff}	Effective volume of test object	[m³]
V_M	Model volume	[m³]
x_M	Model x-position (distance of wheelbase center to nozzle exit plane)	[m]
x_S	X-position of the source that is used to calculate the interference effect	[m]
x_{WE}	X-Position at end of vehicle wake	[m]
x_{MF}	X-Position at end of vehicle model front	[m]

List of Abbreviations

BGK	Bhatnagar-Gross-Krook form
BLPS	Boundary Layer Presuction
CB	Center Belt
CFD	Computational Fluid Dynamics
DBLS	Distributed Boundary Layer Suction
DIVK	Digital model of IVK-MWK
DWT	Standard Digital Wind Tunnel provided by Exa without wind tunnel interference effects (box with low blockage ratio)
Exa	Exa Corporation
ffw	far field wake
FKFS	Research Institute of Automotive Engineering and Vehicle Engines Stuttgart
FWK	IVK Full-scale aero-acoustic wind tunnel
HB	Horizontal buoyancy
IVK	Institute for Combustion Engines and Automotive Engineering, University of Stuttgart
LBM	Lattice-Boltzmann Method
MF	Vehicle Model Front
MWK	IVK model-scale wind tunnel
nfw	near field wake
RH	Ride Height
SAE	Society of Automotive Engineers
SAE-SP	SAE—Special Publication
TBCB	Tangential Blowing in front of the Center Belt
WE	End of vehicle Wake
WRU	Wheel Rotation Unit
NPL	National Physical Laboratory in England

Zusammenfassung

Im aerodynamischen Entwicklungsprozess von Fahrzeugen werden heutzutage verschiedene Entwicklungsmethoden komplementär eingesetzt. Strömungs-simulationen (Computational Fluid Dynamics (CFD)) wurden zunehmend in den existierenden Entwicklungsprozess integriert und haben mittlerweile wesentliche Teile der ursprünglich experimentellen Methoden ersetzt. Diese Koexistenz verschiedener experimenteller und simulativer Methoden erhöht zunehmend die Notwendigkeit eines steten Abgleichs der Ergebnisse beider Methoden.

In den meisten CFD-Simulationen werden die Fahrzeuge in idealisierten Testbedingungen simuliert um den Rechenaufwand der Simulationen zu minimieren. Experimentelle Ergebnisse aus dem Windkanal beinhalten dagegen immer auch sogenannte Interferenzeffekte. Ein direkter Vergleich der Ergebnisse dieser unterschiedlichen Strömungssituationen ist nur bedingt sinnvoll. Die Vergleichbarkeit der Ergebnisse kann z.B. durch die Anwendung entsprechender Korrekturverfahren für Interferenzeffekte auf die experimentellen Werte erreicht werden. In den meisten Fahrzeug-Windkanälen, die für Aerodynamikentwicklung genutzt werden, sind allerdings keine der existierenden Korrekturverfahren im ständigen Einsatz. Mit dieser Arbeit soll die Vergleichbarkeit der Ergebnisse eines empirischen Korrekturverfahrens und der Strömungssimulationen verbessert werden.

Nach einigen einleitenden Worten werden zuerst die beiden wesentlichen Methoden in der aerodynamischen Entwicklung von Fahrzeugen vorgestellt: der Windkanal und die Strömungssimulation (CFD). Im Folgenden wird der Leser in die Grundlagen des Korrekturverfahrens von Mercker und Wiedemann eingeführt. Zu Beginn wird eine kurze Zusammenfassung der historischen Entwicklung und des aktuellen Status des Korrekturverfahrens dargestellt. Danach werden die Grundlagen der Lattice-Boltzmann Methode beschrieben, die der in dieser Arbeit verwendeten CFD-Software zugrunde liegt.

Ein direkter Abgleich zwischen CFD und Experiment ist nur möglich, wenn die Interferenzeffekte auch in der Simulation vorhanden sind. Dazu wurde der

Modellwindkanal des IVK in verschiedenen Windkanalkonfigurationen simuliert, die die statische Druckverteilung in der Windkanalmessstrecke verändern und somit eine Veränderung der wirkenden Interferenzeffekte bewirken. Diese Einflüsse sind direkt in den aerodynamischen Beiwerten des Fahrzeugs sichtbar. Die statische Druckverteilung wurde dabei mit zwei verschiedenen Methoden verändert um auch den Einfluss eines inhomogenen Druckgradienten zu bewerten. Des Weiteren wurden verschiedene Effekte des Windkanals und der Modelle auf die aerodynamischen Beiwerte analysiert. Dabei wurden sowohl Limousinen als auch SUV Modelle in drei verschiedenen Druckgradienten untersucht um einen wesentlichen Einfluss der Fahrzeugform auszuschließen.

Des Weiteren wurde die Anwendung eines Korrekturverfahrens auf die entsprechenden experimentellen Daten aus dem Modellwindkanal des IVK untersucht. Diese Korrekturergebnisse sind direkt mit interferenzfeien Simulationsergebnissen vergleichbar. Abweichungen lassen einen Rückschluss auf die Funktion der Korrekturmethode zu. Basierend auf diesen Erkenntnissen wird in dieser Arbeit eine Änderung der Korrekturmethode vorgeschlagen, die die Vergleichbarkeit von korrigierten experimentellen Ergebnissen und CFD-Simulationen in idealisierten Testbedingungen verbessert. Somit wird in dieser Arbeit eine Verbesserung der untersuchten Korrekturmethode erreicht.

Abstract

The aerodynamic development process of vehicles in the automotive industry today takes advantage of various methods. Computational fluid dynamics (CFD) was integrated into the existing development process, increasing the necessity for comparison between wind tunnel data and CFD simulation results. Most CFD applications simulate idealized test conditions while the wind tunnel data include interference effects due to the wind tunnel geometry. Hence, a comparison of interference-free CFD results with experimental values is of limited use. Comparability can be reached by applying a correction method for wind tunnel interference effects on the measured data. Existing correction methods are not yet widely accepted to be able to utilize for general application in everyday wind tunnel testing. Therefore, the goal of this thesis is to gain a deeper insight into the comparability between an empirical correction method for open-jet interference effects and CFD results.

After some introductory remarks on motivation and challenges for this thesis, the two main aerodynamic development tools are described: wind tunnel and CFD. Wind tunnel interference effects, and the correction method proposed by Mercker and Wiedemann, are introduced. A brief summary of the historical development process and the actual status of this correction method are outlined.

Thereafter, the aerodynamic CFD model, using the Lattice-Boltzmann method (LBM) for simulating vehicles inside the "Institute for Combustion Engines and Automotive Engineering, University of Stuttgart" (IVK) model-scale wind tunnel test section, is depicted.

A direct comparison between CFD and experiment is only possible when the interference effects are also simulated. As such, the IVK model wind tunnel is completely simulated in diverse configurations to change the pressure distribution in the wind tunnel test section, thereby changing the interference effects. The different static pressure distributions result in changed aerodynamic drag due to interference between the wind tunnel geometry and the pressure on the surface of the vehicle. The wind tunnel pressure distribution is modified with two different methods to research the influence of inhomogeneous pressure distributions. Various effects of the wind tunnel and model setup on the aerodynamic drag and lift are investigated. The effects of three different wind tunnel pressure distributions are shown for sedan and SUV models, presenting the interaction between vehicle shape and wind tunnel interference effects.

Second, the possibility to apply a standard wind tunnel correction method to the experimental data from the IVK model wind tunnel is examined. These correction results are comparable with simulation results with negligible blockage effects. From discrepancies between these results, conclusions are made concerning the functionality of the correction method. Furthermore, changes in the correction method are recommended in this thesis to improve the comparability between wind tunnel data and CFD simulation results. Therefore, the result of this thesis is an improvement in the correction method.

1 Introduction

The influence of aerodynamics on several vehicle functions, such as vehicle performance, fuel consumption, stability, thermal management, comfort, and safety, shows the interdisciplinary character of vehicle aerodynamics in the vehicle development process. The importance of aerodynamics has even increased lately, due to stricter legislation concerning fuel consumption and emissions.

1.1 Aerodynamic Vehicle Development – Experiment and Simulation

A variety of development methods are used today in the aerodynamic development process of vehicles in the automotive industry. Until the end of the 1960s, road tests and wind tunnel tests have been the only applied methods used in vehicle development history [1].

Since the 1980s, computational fluid dynamics (CFD) was increasingly integrated into the aerodynamic design process of vehicles in the automotive industry. But, it is unlikely that CFD will be able to replace the wind tunnel completely in the near future [2]. Therefore, it is becoming more necessary to improve the conformity of wind tunnel data and results from CFD simulations within the development process.

The comparability of test results obtained by these three different methods should be guaranteed by equal test conditions. In aerodynamics, idealized test conditions are used, which are not representative of typical flow conditions that a vehicle experiences on road. These real road conditions feature flow structures caused by natural stochastic wind, gustiness, or variations in space and time [3]. The idealized conditions used for the aerodynamic optimisation are a free stream of nearly infinite extension, ground movement relative to the vehicle, wheel rotation, no side wind, and no turbulence. Turbulence levels

© Springer Fachmedien Wiesbaden GmbH, part of Springer Nature 2018
O. Fischer, *Investigation of Correction Methods for Interference
Effects in Open-Jet Wind Tunnels*, Wissenschaftliche Reihe
Fahrzeugtechnik Universität Stuttgart, https://doi.org/10.1007/978-3-658-21379-4_1

and turbulent length scales, either caused by natural wind or other vehicles, are completely neglected.

A wind tunnel also attempts to simulate these idealized conditions, but fails to do so to some degree due to interference effects, even if wheel rotation and ground simulation are applied. These interference effects originate from the limited dimensions of the wind tunnel geometry, having a significant influence on measurements in all wind tunnels.

In most CFD simulations, the vehicle is simulated under the idealized test conditions described above. Hence, the influence of wind tunnel interference effects is neglected, and the comparability of interference-free CFD results with experimental values is restricted.

There is also the possibility of correction methods directly applied to the measurement results in the wind tunnel to reduce the influence of interference effects. Wind tunnel interference effects and possible analytical correction methods have been investigated for several decades, with several papers published in recent years addressing this subject [4]-[16]. An excellent summary of blockage effects in wind tunnels was edited by Ewald [17]. Wind tunnel corrections are mandatory in closed-wall wind tunnels, having been developed and applied for decades. However, interference effects in open-jet wind tunnels are still an object of research because no existing correction method is widely accepted enough to be used for general application in everyday wind tunnel testing.

All of these problems lead to the necessity of a better overall understanding of wind tunnel interference effects.

1.2   Motivation and Problem

Basically, three motivations exist for increasing the overall understanding of wind tunnel interference effects.

1.   The comparability of measurement results from different wind tunnels is still considered unsatisfactory. Due to wind tunnel interference effects

and ground simulation effects, there are barely two wind tunnels that provide the same results for a vehicle measured in the same configuration. This necessitates conducting the entire development process of one vehicle in one facility only. Furthermore, many of the of benchmark measurements of competitor vehicles, which are still conducted today, could be reduced.

2. The conformity of CFD results and experimental results is still problematic, as interference effects are generally neglected in CFD simulations (see Chapter 1.1).

3. The growing number of wind tunnel applications in the industry is the reason for new automotive wind tunnels to still be built today. These wind tunnels generate considerable construction and operating costs; therefore, each new wind tunnel faces a conflict. On the one hand, limited budget and construction area demand a small wind tunnel. On the other hand, the wind tunnel size must be sufficient to reduce interference effects to an acceptable extent. A better understanding of wind tunnel interference effects could help reduce the necessary size and, thereby, costs of new wind tunnels.

Hence, it is the goal of this thesis to gain a deeper insight into interference effects and improve the empirical correction method for open-jet interference effects. The main focus lies on the horizontal buoyancy effect, which often is the strongest of all open-jet interference effects.

CFD is the most adept tool to investigate these effects as it provides the whole flow field information, ensuring that all effects are captured correctly in a simulation of the wind tunnel test section with a vehicle model present. Therefore, a detailed digital wind tunnel model must be developed to enable a comparison of CFD and experimental results, making it possible to (a) understand the significance of various details of the test environment, and (b) gain a better understanding of wind tunnel interference effects. This detailed wind tunnel model needs extensive validation between simulation and experimentation. The "Institute for Combustion Engines and Automotive Engineering, University of Stuttgart" (IVK) model-scale wind tunnel (IVK-MWK) was chosen due to various known methods to change the static pressure distribution in the test section.

All simulations are conducted using Exa PowerFLOW, which was chosen because of its transient nature due to the underlying Lattice-Boltzmann algorithm. It is capable of capturing all turbulence effects in the highly turbulent shear layer. The shear layer development directly affects the static pressure distribution along the test section and, therefore, is important for a correct simulation of interference effects.

2 Aerodynamic Development Tools

In general, three methods are used today in the aerodynamic development of vehicles in the automotive industry. The dominating method is the aerodynamic measurement in a wind tunnel. Second is the road experiment method, which is used infrequently, as compared to the wind tunnel. CFD is the third method. It has gained importance in the last twenty years and is considered to be an important tool for an aerodynamicist, besides the wind tunnel. Today, CFD is considered a valuable developmental tool in the aerodynamic development process. These three methods offer several advantages and disadvantages.

It is obvious that flow conditions, while driving on road, are closest to reality. But, it is impossible to create reproducible laboratory conditions due to the additional varying effects of wind, weather, turbulence, and air temperature. Furthermore, high accuracy measurements of aerodynamic forces and acoustics, in a moving vehicle, are difficult, or even impossible, to conduct.

Therefore, wind tunnel and CFD are the symbiotic tools of choice for aerodynamic development today. The advantages and disadvantages are discussed in this chapter.

2.1 Wind Tunnel Technology

The wind tunnel provides the required reproducible test conditions and short turnaround times. Force measurements can easily be produced, without major changes on the vehicle, by using an external balance. Wind tunnels, with appropriate equipment, offer the possibility of easy aero-acoustic measurements around and inside the vehicle. The disadvantages are (a) high acquisition and maintenance costs, (b) difficulties with the simulation of rolling road conditions including rotating wheels, and (c) the lack of comparability between different wind tunnels.

© Springer Fachmedien Wiesbaden GmbH, part of Springer Nature 2018
O. Fischer, *Investigation of Correction Methods for Interference
Effects in Open-Jet Wind Tunnels*, Wissenschaftliche Reihe
Fahrzeugtechnik Universität Stuttgart, https://doi.org/10.1007/978-3-658-21379-4_2

The basics of wind tunnel technology, upon which this thesis is grounded, are depicted within Chapter 2.1.1.

2.1.1   Wind Tunnel Design

2.1.1.1   Wind Tunnel Types

Numerous different designs for subsonic wind tunnels have been developed and there are almost endless variations of the specific features of various tunnels. Virtually, every wind tunnel is one of a kind [18]. Two general design attributes for subsonic wind tunnels can be identified which are, (a) the basic type of wind tunnel, and (b) the basic type of test section configuration.

These two wind tunnel types are open circuit and closed circuit. An open circuit wind tunnel is constructed of an air intake, contraction, test section, diffuser, fan section, and air exhaust (see Figure 2.1). The surrounding air enters the wind tunnel flow path at the intake and leaves the ducting at the exhaust. The tunnel may have various test section types, as long as the plenum surrounding the test section is not ventilated to the exterior. The whole tunnel can be built-in open air or within a closed building.

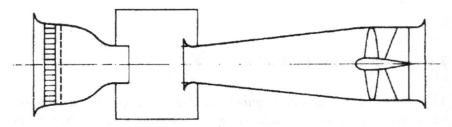

Figure 2.1:   Open circuit wind tunnel with an open-jet test section ("Eiffel") [19]

A closed circuit wind tunnel consists of the same parts as that of an open circuit wind tunnel, but with an additional ducting closing the circuit (see Figure 2.2). The air flowing in a closed circuit wind tunnel recirculates continuously, with little or no exchange of air with the exterior. The majority of closed circuit wind tunnels have a single return duct, although tunnels with double returns have been built. The closed circuit tunnel may have various test section types [18].

In a return-flow wind tunnel, the return duct must be properly designed to reduce the pressure losses and to ensure low turbulence in the test section. The ducting of the wind tunnel is designed to reduce flow separation and to maximize cross-section at each part to reduce air velocity and, thereby, reduce pressure losses.

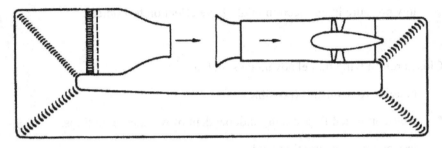

Figure 2.2: Closed circuit wind tunnel with an open-jet test section ("Göttingen") [19]

There are advantages and disadvantages for both wind tunnel types. When a purchasing decision concerning a new wind tunnel is to be made, the funds available and the purpose are the general criteria used in the decision-making process [18]. Further advantages and disadvantages are listed below [18].

Open return wind tunnel advantages include,

■ low construction costs, and

■ possibility of running internal combustion engines or extensive visualization with smoke.

Open return wind tunnel disadvantages include,

■   no aero-acoustic measurements possible, as high noise levels,

■   more energy required to run,

■   inlet and exhaust, if open to the atmosphere, can cause weather to affect the flow conditions in the wind tunnel, or objects to be blown into the air intake, and

■   low pressure in test section might have effect on test object.

Closed return wind tunnel advantages include,

■   better aero-acoustics (possible),

■   well-controlled flow quality independent of weather conditions,

■   less energy consumption, and

■   less environmental noise.

Closed return wind tunnel disadvantages include,

■   high construction costs,

■   necessity of air cooling, in most cases, resulting in higher pressure losses and, therefore, higher energy consumption,

■   higher costs for building construction, and

■   necessity for wind tunnel purges.

2.1.1.2 Wind Tunnel Test Sections

The two basic test section configurations are open-jet test section and closed-walls test section (see Figure 2.3). These configurations are considered to be the two ends of a spectrum of test section designs, including also slotted, streamlined, and adaptive walls [18].

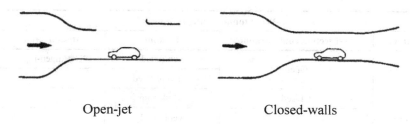

Open-jet   Closed-walls

Figure 2.3: Wind tunnel test section types (open-jet, closed-walls) [1]

The main reasons for the occurrence of the change from closed test sections to open-jet test sections during the last two decades, are associated with aero-acoustics (semi-anechoic test section) and test section accessibility. The latter is important to be able to apply configuration changes of the test object quickly and efficiently. However, in dedicated aerodynamic tunnels for race cars, closed test sections still prevail, because the same model is tested almost all the time and blockage effects, especially due to wing-upwash, are deemed to be corrected more precisely.

2.1.1.3   *Specific wind tunnel configurations*

Only a small number of combinations of wind tunnel types and test section configurations are possible. These combinations are shown in Table 2.1.

Table 2.1:   Possible combinations of wind tunnel types and test section types

Wind tunnel configurations			
Circuit	**Test section configuration**	**Plenum**	**Design name**
Closed	Open	ventilated	"Göttingen-type"
		not ventilated	
	Closed	no plenum	
	slotted walls	ventilated	
		not ventilated	
	streamlined walls	no plenum	
	adaptive walls	no plenum	
Open	Open	not ventilated	"Eiffel type"
	Closed	no plenum	"NPL"
	slotted walls	not ventilated	
	streamlined walls	no plenum	
	adaptive walls	no plenum	

Some combinations of wind tunnel type and test section design have been given specific names (see Table 2.1). Open circuit wind tunnels with an open-jet test section are called "Eiffel type" wind tunnels. Open circuit wind tunnels with a closed-wall test section are referred to as "NPL" wind tunnels (after the National Physical Laboratory in England). The "Göttingen-type," which was designed by Ludwig Prandtl, features air circulation in a closed duct system and an open-jet test section. Most automotive wind tunnels built today are Göttingen-type wind tunnels.

2.1.2 Wind Tunnel Calibration

The determination of reference dynamic pressure is of fundamental importance because it is used for all dimensionless quantities derived from wind tunnel measurements. The reference velocity and, consequently, the tunnel speed are usually defined as the average flow velocity in the empty wind tunnel in the region of the test section, which will be occupied by the test object [1]. In the empty test section, the velocity can be measured with a flow probe (i.e., a pitot static tube) [1]. When a test object is placed in the test section, the reference velocity, or the dynamic pressure, in the test section can no longer be measured by a flow probe. Therefore, it must be determined from a calibration procedure.

A pitot static tube is used for the calibration procedure in the empty test section. It is located in the region of the test section where the test body will be placed; and the dynamic pressure is calculated as the difference between measured total pressure and static pressure at the pitot static tube according to the equation of Bernoulli.

$$\text{Bernoulli's equation: } p_{tot} = p_{stat} + q = p_{stat} + \frac{\rho}{2}v^2 = const. \qquad \text{Eq. 2.1}$$

The dynamic pressure q can be calculated by:

$$q = \frac{\rho}{2}v^2 \qquad q = p_{tot} - p_{stat} \qquad \text{Eq. 2.2}$$

p_{tot}	:	total pressure
p_{stat}	:	static pressure
q	:	dynamic pressure
v	:	Velocity

Typically, the dynamic pressure is measured at several positions for a wind tunnel calibration, and the results are averaged over space.

The mass flow can be measured by a static pressure difference Δp between two different cross-sectional areas of the wind tunnel duct. The biggest differences of cross-sectional areas and, therefore, pressure differences, are found at the nozzle contraction. In automotive open-jet wind tunnels, two methods are used (see Figure 2.4) [5][20]:

- the nozzle method, and

- the plenum method.

The pressure difference Δp_N of the nozzle method is measured between two cross-sections within the nozzle contraction, such as the settling chamber (p_{SC}), and a cross-section close to the nozzle exit plane (p_N). The plenum method uses the pressure difference Δp_P between settling chamber (p_{SC}) and plenum (p_P) (see Figure 2.4).

$$\Delta p_N = p_{SC} - p_N \qquad \Delta p_P = p_{SC} - p_P \qquad \text{Eq. 2.3}$$

p_{SC} :   pressure in wind tunnel settling chamber

p_N :   pressure in the nozzle contraction close to nozzle exit plane

p_P :   pressure in the wind tunnel plenum

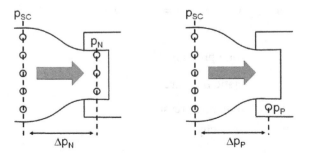

Figure 2.4:   Determination of dynamic pressure in the wind tunnel test section with nozzle method (left) and plenum method (right)

For the wind tunnel calibration, the measured pressure differences of nozzle method (Δp_N) or plenum method (Δp_P) are multiplied by a calibration factor determined in the empty tunnel to obtain the dynamic pressure q.

$$q = p_{tot} - p_{stat} = \frac{\rho}{2} v^2 = k \cdot \Delta p$$

$$q = k_N \cdot \Delta p_N \qquad\qquad \text{Eq. 2.4}$$

$$q = k_P \cdot \Delta p_P$$

k :   nozzle factor

k_N :   nozzle factor for the nozzle method

k_P :   nozzle factor for the plenum method

The relation between q and Δp is approximated to be linear. A better approximation can be found using a polynomial. The reason for the non-linear behavior of k is observed in the Reynolds number behavior of the nozzle wall boundary layer [5].

The applicability of both methods depends on the blockage condition in the respective wind tunnel. The blockage condition is determined by the basic dimensions of wind tunnel test section and tested model.

The ratio of these dimensions is a measure for the blockage condition in a wind tunnel test section. The flow field around a test object will almost not be distorted in the boundary areas (i.e., close to walls or shear layer) when the test object is small as compared to the test section size. This situation shall be referred to as a low blockage condition. For bigger test objects, which significantly change the flow field also in the boundary areas, the situation in a wind tunnel test section is referred to as high-blockage condition.

A mathematically-defined measure for the blockage condition is the blockage ratio (see Chapter 0). For low blockage conditions, both determination methods deliver the same results. In this instance, the situation corresponds closely to the calibration case, where only the calibration probe is present in the test section [21]. In high-blockage cases, the adverse pressure gradient in front of the model may extend into the nozzle and affect the measured nozzle pressure (p_N). As long as this is not the case, the nozzle method produces a constant volume flow through the nozzle, which means the empty-tunnel volume flow will be sustained even with the vehicle placed in the test section.

In most open-jet and slotted wall wind tunnels, a duct system connects the plenum chamber with the surrounding atmosphere. This leads to plenum pressure being equal to the atmospheric pressure surrounding the wind tunnel. Therefore, as the plenum pressure will remain constant, the plenum method is recommended for high-blockage configurations. Obviously, a plenum chamber is needed for the plenum method, therefore, it can only be used for open-jet or slotted wall configurations. Nevertheless, both methods need a correction, as the presence of the model influences the nozzle flow to some extent [5].

The direct measurement by a pitot static tube located inside the flow is one of alternative possibilities for the measurement of the reference velocity. This method - though it may be used in some wind tunnels - is not recommended due to the expectation of larger errors, as compared to the methods described above, when a model is present in the test section.

2.2   Computational Fluid Dynamics - Numerical Scheme

One advantage of a CFD simulation is the complete set of information of the flow field which each simulation delivers. This offers deeper insight into flow structures and helps facilitate understanding of the effects of geometry changes. Comparable experimental results would require time consuming and expensive wind tunnel measurements. Furthermore, results of standard CFD simulations are not influenced by wind tunnel interference effects [22].

The turnaround time, as compared to wind tunnel experiments, is the remaining disadvantage. But the influence of computational simulations on the vehicle development process has grown constantly with the increased computing power and simulation accuracy [22].

Within this thesis, CFD is used to offer deeper insight into the flow field of an open-jet wind tunnel. Moreover, it is used to generate results, which are free of wind tunnel interference effects. All CFD simulations are performed using Exa's PowerFLOW, a Lattice-Boltzmann solver. More details about the Lattice-Boltzmann method (LBM) and the additional models used by Power-FLOW are to follow [22].

2.2.1   The Lattice-Boltzmann Method

Unlike traditional CFD techniques that consider the fluid at a macroscopic level, the LBM is based on the mesoscopic Lattice-Boltzmann equation, describing the evolution of a discrete particle distribution function f_i, which characterizes the number of particles at a time t, with a momentum p at each node of a spatial grid (lattice) [22]. The Lattice-Boltzmann equation has the following form:

$$f_i\left(\vec{x}+\vec{c}_i\Delta t,t+\Delta t\right)-f_i\left(\vec{x},t\right)=C_i\left(\vec{x},t\right) \qquad \text{Eq. 2.5}$$

where f_i is the particle distribution function moving in the ith direction, according to a finite set of the discrete velocity vectors $\{\vec{c}_i : i=0, ...b\}$. $\vec{c}_i \cdot \Delta t$ and Δt are space and time increments, respectively.

The right hand side of equation Eq. 2.5 is known as the collision term and obeys the basic conservation laws of mass and momentum. The Bhatnagar-Gross-Krook (BGK) form described in [22][23][24][25][26] is used in Power-FLOW:

$$C_i\left(\vec{x},t\right)=-\frac{1}{\tau_r}\left[f_i\left(\vec{x},t\right)-f_i^{eq}\left(\vec{x},t\right)\right] \qquad \text{Eq. 2.6}$$

where f_i^{eq} is the local equilibrium distribution function, which depends on local hydrodynamic properties and τ_r is the relaxation time parameter which is linked to the kinematic viscosity by [22][25][26][27][28]:

$$v = \left(\tau_r - 1/2\right)T \qquad \text{Eq. 2.7}$$

The set of equation from Eq. 2.5 to Eq. 2.5 can be solved for f_i, and the basic macroscopic quantities of the system can then be recovered through a simple moment summation:

$$\rho\left(\vec{x},t\right)=\sum_i f_i\left(\vec{x},t\right), \quad \rho\vec{u}\left(\vec{x},t\right)=\sum_i \vec{c_i} f_i\left(\vec{x},t\right) \qquad \text{Eq. 2.8}$$

The combination of equation Eq. 2.5 to Eq. 2.8 forms the typical LBM scheme for fluid dynamics. The transient compressible Navier-Stokes equations can be recovered in the low Mach number limit under certain conditions [22][25][26][27][28].

2.2.2   Fluid Turbulence Model and Wall Model

The molecular relaxation time τ_r in the collision term Eq. 2.6 can be modified to account for the turbulent fluctuations [29]. This is a major difference from the classical Reynolds-averaged Navier-Stokes technique that usually couples k and ε to the momentum equation through a linear relation to the Reynolds stress term (Bousinnesq approximation) that eliminates any nonlinearity of the Reynolds stress. By avoiding this approach, the LBM will contain the higher order terms [22].

The modification is done by replacing the molecular relaxation time scale with an effective turbulent relaxation time scale τ_{eff} derived from a systematic renormalization group procedure [30] as:

$$\tau_{eff} = \tau_r + C_\mu \frac{k^2/\varepsilon}{T\left(1+\widetilde{\eta}^2\right)^{1/2}} \qquad \text{Eq. 2.9}$$

where $\tilde{\eta}$ is a combination of a local strain parameter $\eta = k|S|/\varepsilon$, local vorticity parameter $\eta_\omega = k|\Omega|/\varepsilon$, and local helicity parameters. k and ε are the turbulent kinetic energy and the rate of kinetic energy dissipation. S is the strain rate tensor. Ω is the vorticity [22].

The sub-grid scale energy dissipation is modeled by a modified k-ε two-equation model based on the original renormalization group formulation [22][31][32][33]:

$$\rho\frac{Dk}{Dt} = \frac{\partial}{\partial x_j}\left[\left(\frac{\rho v_0}{\sigma_{k_0}} + \frac{\rho v_T}{\sigma_{k_T}}\right)\frac{\partial k}{\partial x_j}\right] + \tau_{ij}S_{ij} - \rho\varepsilon \qquad \text{Eq. 2.10}$$

$$\rho\frac{D\varepsilon}{Dt} = \frac{\partial}{\partial x_j}\left[\left(\frac{\rho v_0}{\sigma_{\varepsilon_0}} + \frac{\rho v_T}{\sigma_{\varepsilon_T}}\right)\frac{\partial \varepsilon}{\partial x_j}\right] + C_{\varepsilon_1}\frac{\varepsilon}{k}\tau_{ij}S_{ij} - \left[C_{\varepsilon_2} + C_\mu\frac{\tilde{\eta}^3(1-\tilde{\eta}/\eta_0)}{1+\beta\tilde{\eta}^3}\right]\rho\frac{\varepsilon^2}{k} \qquad \text{Eq. 2.11}$$

The eddy viscosity is defined with the classical formulation:

$$v_T = C_\mu k^2/\varepsilon \qquad \text{Eq. 2.12}$$

A modified explicit Lax-Wendroff time marching finite difference scheme is used to solve equations Eq. 2.10 and Eq. 2.11 on an original LBM grid. In addition to the turbulence model, a turbulent wall model is used to provide approximate boundary conditions for the near wall particles. The model is based on the traditional Spalding's law of wall [34][35] and is extended to include pressure gradient sensitivity [22][36].

3 Wind Tunnel Interference Effects

Wind tunnels attempt to simulate the flow around a test object, which is located within an infinite stream. Primarily due to limited financial resources, the cross-section of the nozzle and the size of the test section are limited. The proximity of the surrounding wind tunnel geometry leads to a different flow field around the test object than within an infinite stream. These effects get stronger with increasing blockage ratio B, which is defined as the ratio of the frontal areas of the test object and the nozzle cross-section.

The blockage ratio B is defined as:

$$B = \frac{A_M}{A_N}$$

Eq. 3.1

A_M :   frontal area of vehicle model

A_N :   nozzle cross-section

The collector blockage ratio B_C can be defined respectively as:

$$B_C = \frac{A_M}{A_C}$$

Eq. 3.2

A_C :   collector cross-section

Responsible for the constrained flow field, are the wind tunnel interference effects. Ground simulation effects, not included within wind tunnel interference effects, are due to the moving road relative to the vehicle and turning wheels. Wind tunnel interference effects have been investigated for several decades. In closed-wall wind tunnels, they could not be considered negligible, which led to the development of both other test section designs (see Chapter 2.1.1), and suitable correction methods. Therefore, the development of closed-wall correction methods has a long history. Today, these correction methods

© Springer Fachmedien Wiesbaden GmbH, part of Springer Nature 2018
O. Fischer, *Investigation of Correction Methods for Interference Effects in Open-Jet Wind Tunnels*, Wissenschaftliche Reihe Fahrzeugtechnik Universität Stuttgart, https://doi.org/10.1007/978-3-658-21379-4_3

are used in everyday aerodynamic practices in most closed-wall wind tunnels [37].

While wind tunnel interference effects in open-jet wind tunnels have been neglected, today it is widely accepted that the limited extent of the test section in an open-jet wind tunnel causes various interference effects which have a considerable impact on the measured aerodynamic forces [4][11]. A non-uniform static pressure distribution along the test section especially generates additional forces on the test object due to buoyancy. Correction methods for open-jet interference effects are still discussed avidly and widely within the community. In most wind tunnels, existing correction methods are not used on a regular basis.

3.1   Blockage Effects in Open-Jet Wind Tunnels

Mercker and Wiedemann proposed a correction method based on five different interference effects which occur in open-jet wind tunnels [4]. The correction of four out of five of these interference effects is a correction of dynamic pressure q and, therefore, of wind velocity seen by the vehicle. The fifth interference effect is horizontal buoyancy, which is due to the static pressure distribution present in most open-jet wind tunnel test sections. The classic correction for horizontal buoyancy is done by a calculation of the horizontal buoyancy force acting on the vehicle. The calculation of the horizontal correction term has been extensively discussed since the 1990s due to the static pressure distribution influencing the vehicle's near field wake and thereby the measured drag [6][10][11].

The different interference effects are illustrated in Figure 3. and shall be described briefly in the following paragraphs.

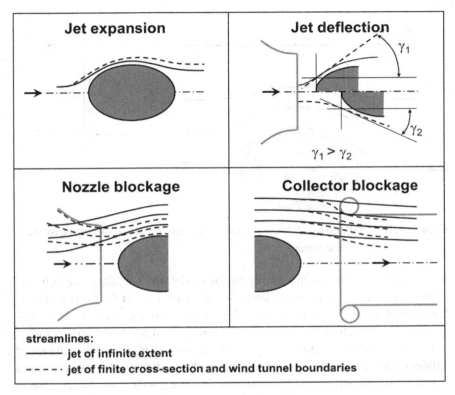

Figure 3.1: Blockage effects in open-jet wind [1]

3.1.1 Jet Expansion Effect

A jet of finite cross-section behaves differently, as compared to a jet of infinite extent, when a model is present inside the jet, for there are different boundary conditions in both cases. In a jet with finite cross-section, the static pressure along the boundary streamline is constant and equal to the static pressure of the ventilated plenum chamber. In an infinite jet, this ambient static pressure is found in an infinitely large distance from the model, as shown in Figure 6.

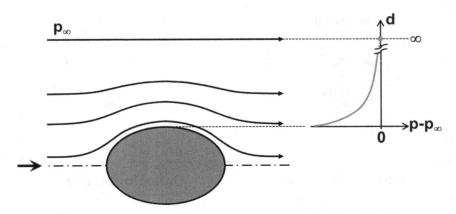

Figure 3.2:   Undisturbed flow with ambient static pressure in large distance
from the model

When the flow is displaced laterally by the model, the boundary streamline is
curved convexly. Due to the different boundary conditions for finite jet cross-
section and infinite jet cross-section, the curvatures of the streamlines are different. Boundary streamlines in a jet with finite extent do have a smaller radius
of curvature than streamlines in a jet with infinite extent, due to the balance of
ambient static pressure and boundary streamline static pressure. An increased
curvature of streamlines is accomplished by an expansion of the jet. Due to
continuity, this results in lower average velocity around the vehicle, and thus,
lower measured drag. This effect is also called "solid-blockage."

3.1.2   Nozzle blockage effect

The distance between nozzle exit plane and model is crucial for "nozzle blockage". The stagnation region in front of the model may extend into the nozzle,
which generates a displacement of the flow, resulting in a modified velocity
profile in the nozzle exit plane. The generated velocities, thereby, result in
altered forces acting on the vehicle. The method of q determination (nozzle
method or plenum method, see 2.1.2) used in the wind tunnel is also essential
for the modified velocity profile.

3.1.3 Jet Deflection Effect

The jet may be deflected by the vehicle due to proximity of nozzle and vehicle which leads to an expansion of the jet. The dependency of distance and deflection angle is shown in Figure 3.. This effect is in addition to the above already mentioned jet expansion. In both effects, jet expansion causes lower velocities around the vehicle and, therefore, lower measured drag forces.

3.1.4 Collector Effect

The collector effect denotes the interference effect of the vehicle's wake with the flow entering the wind tunnel collector. The flow is displaced by the wake of the vehicle, and the resulting interference is a similar effect to the nozzle blockage effect. The vehicle is affected by this interference with the collector walls, comparable to the effect in a closed-wall wind tunnel.

3.2 Effect of Static Pressure Distribution (Horizontal Buoyancy)

In all open-jet wind tunnels, a static pressure distribution exists along the test section length. This pressure distribution is generated by several mechanisms. Both nozzle design and collector design influence the distribution significantly. Also, the shear layer, which develops between free jet and still air in the plenum chamber, influences the static pressure distribution.

Shear layer development, and thereby static pressure distribution, can be strongly influenced by vortex generators in the nozzle exit plane, like Seiferth- or delta-wings. In wind tunnels, such vortex generators are generally used to prevent wind tunnel buffeting by destroying large-scale vortices in the shear layer.

The influence of collector design on static pressure distribution is due to the flow deceleration in front of the collector. Therefore, the static pressure distribution can be widely influenced by changing collector design or collector ventilation through breathers. Collector design in aero-acoustic wind tunnels is always a compromise between aerodynamics and aero-acoustic performance. The collector can only rarely be optimized for just an ideal static pressure distribution, with the consequence that various static pressure distributions can be found in existing open-jet wind tunnels.

The static pressure distribution generates an additional drag force on the vehicle due to the static pressure difference between vehicle front and vehicle rear (Figure 3.3). This horizontal buoyancy effect—where the pressure gradient is mostly positive—usually reduces the vehicle drag. In correction methods, horizontal buoyancy is often the most important correction component, as it generates the largest drag changes in most open-jet wind tunnels.

Moreover, there has been a vital discussion in recent years concerning the influence of a static pressure distribution on the vehicle's near field wake and, thereby, on the measured drag. New approaches to account for this wake effect were published by Wickern [6] and Mercker et al. [10][11].

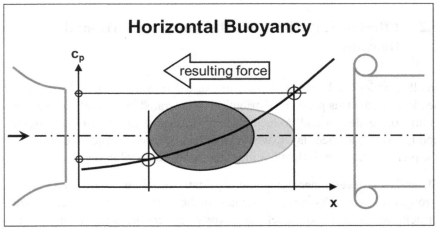

Figure 3.3:   Horizontal buoyancy in open-jet wind tunnels [1]

4 Correction Method for Interference Effects in Open-Jet Wind Tunnels

In contrast to wind tunnel corrections in closed-wall wind tunnels, the necessity of corrections of open-jet wind tunnel measurements for automotive applications has been a topic of discussion as they are significantly smaller [4].

Boundary conditions for closed test sections are precise, while boundary conditions for a free jet are only approximate [4]. Furthermore, closed test sections are considerably longer than open-jet test sections, where the model is positioned much closer to the nozzle and collector. These significant differences have impeded the development of a correction method for open-jet tunnels compared to closed-wall wind tunnels.

Corrections for wind tunnel interference effects must be divided into two correction categories: corrections to the dynamic pressure and corrections because of static pressure gradients.

It is assumed, that most interference effects lead to a flow situation, where the model experiences a velocity different from the velocity measured by the wind tunnel control system. The simplest example for this is a closed-wall wind tunnel where the test object will experience a higher velocity compared to the empty test section due to continuity, if the volume flux is constant. The classical correction methods for open and closed wind tunnel test sections, therefore, correct the dynamic pressure. The correction is calculated from the displacement of the test object and a correction of the angle of attack. The displacement of the test object is modeled using a combination of potential-flow source and sink singularities [6].

A static pressure distribution basically exists in all kinds of test sections, but primarily, it is present in open-jet wind tunnels, due to the stagnation of the flow in front of the collector. The resulting effect can be described as similar to a vertical pressure gradient in a dense fluid, which leads to a buoyancy force as the test object experiences a varying static pressure from nose to tail. This effect is known as horizontal buoyancy due to the direction of the force parallel to the tunnel symmetry axis, which is usually horizontal. In today's automotive

© Springer Fachmedien Wiesbaden GmbH, part of Springer Nature 2018
O. Fischer, *Investigation of Correction Methods for Interference
Effects in Open-Jet Wind Tunnels*, Wissenschaftliche Reihe
Fahrzeugtechnik Universität Stuttgart, https://doi.org/10.1007/978-3-658-21379-4_4

open-jet wind tunnels, horizontal buoyancy can be identified as the interference effect, which affects the measured forces the most.

Today, most automotive open-jet tunnels still do not use correction methods, although testing is done with considerable blockage ratios. One reason for this could be that interference effects in open-jet wind tunnels are still an object of research. There is no existing correction method today that is widely accepted and ready for use in general application in everyday wind tunnel testing. Possible open-jet correction methods have been investigated and several papers have been published examining this subject. A reason for the missing acceptance of correction methods could be that the correction usually increases the aerodynamic drag coefficient. Car companies seek to lower the aerodynamic drag coefficient with great effort in order to develop vehicles that are more fuel efficient. Therefore, rising coefficients due to correction methods could cause communication problems with the public, and also within these companies.

An ideal wind tunnel correction method should be applicable without extensive numerical calculations or other time consuming requirements, like additional measurements. These circumstances would guarantee the correction method to be applicable as a fast online tool to calculate corrected forces directly after the measurement. This would require that highly simplified flow models be used to approximate the test section interference. But, improved computational facilities in the future may permit more complex flow models, such as panel methods or other approaches of similar complexity [21].

4.1   Historical Development

The correction method for open-jet test sections used in this thesis is the latest version of a correction method which has evolved over several years, originating with a paper by Mercker and Wiedemann [3]. The Mercker-Wiedemann method expressed open-jet blockage as a sum of four effects (for the first time):

- jet expansion,

- jet deflection,

- nozzle blockage, and

- collector blockage,

Which, when added with horizontal buoyancy, results in five interference effects. Before the advent of the Mercker-Wiedemann method, jet expansion and horizontal buoyancy were considered to be the only interference effects to occur in open-jet wind tunnels. See examples in [18][38].

The open-jet correction method, used in this thesis, has emerged from the original Mercker-Wiedemann method. The development process is summarized in the following chapter and has also been documented in various SAE-papers [4][5][10][11].

4.1.1   Open-Jet Blockage Effects

The correction for jet expansion has been the first part of an approach for open-jet correction. It is based on Lock's interference correction, which uses mirror images in potential-flow theory to represent wind tunnel boundaries [39]. Mercker used this approach, and a publication by Young and Squire [40], to develop a solid-blockage correction in 1986 [41], for a closed-wall tunnel. Mercker suggests using the vehicle volume as a measure for solid-blockage ("volume method"). The SAE special publication (SAE-SP) on closed test section blockage corrections by Cooper [37], compared several correction methods and discovered that Mercker's method for closed-wall test sections delivered the best results, as well as two other methods. Among these three methods, Mercker's method has the advantage, in that it needs no extra measurements in contrast to the other two methods, which require additional pressure measurements on the test section walls. As the method contains empirically determined constants, it must be restricted to the vehicle classes of vehicle shapes used in determining these constants [37]. The first SAE-SP on open-jet wind tunnel corrections, by Buchheim et al. [42], mainly documented the blockage effects observed in a large number of benchmark cases; and, it also mentioned Mercker's approach.

4.1.2   Horizontal Buoyancy

As a first approach, the additional force generated by a static pressure gradient was calculated by a simple buoyancy calculation [4]:

$$F = \int p dA = \int \frac{\partial p}{\partial x} dV$$

Eq. 4.1

A :   surface area of test object

V :   volume of test object

Munk [43] showed that this simple calculation is not valid for bodies in moving fluid and proposed to use a certain effective volume V_{eff}, which is greater than the actual body volume [4]. Later, Glauert identified the effective volume of a sphere and a circular cylinder to be 1.5 and 2 times the actual body volume [44]. This factor was later called the Glauert Factor.

$$V_{eff} = G \cdot V$$

Eq. 4.2

G :   Glauert Factor

V_{eff} :   effective volume of test object

4.1.3   Mercker-Wiedemann Method

Mercker's closed-wall correction [41] became the basis for a semi-empirical open-jet correction method, published in 1996 by Mercker and Wiedemann [4]. For the first time, a correction method introduced additional blockage effects due to the proximity of the fixed walls of nozzle and collector [6]. Their correction method identified five interference effects for an open-jet test section, including classical jet expansion correction. The test section interference, in open-jet test sections, is subdivided into a proportion, (a) due to the effect of the open-jet of infinite length, and (b) due to the impermeable walls of the nozzle in the proximity of the model and a similar part due to the presence of the collector. For the collector interaction, a semi-empirical model has been proposed which is based on the closed-wall approach, but is still a part of the standard correction procedure [41]. The fifth effect is horizontal buoyancy, which is due to the static pressure distribution along the open test section.

While the first version [4] was developed only for wind tunnel measurements performed with the nozzle method, it was expanded to measurements done with the plenum method one year later [5]. As such, the Mercker-Wiedemann method became able to manage data from both kinds of open-jet test sections (i.e., those which use the nozzle method, thus keeping the volume flow inside the test section constant; and tunnels using the plenum method and sustaining U_∞ at the nozzle boundary for all blockage conditions). The latter leads to lower velocities and, therefore, lower forces than for the nozzle method [5].

For the correction of horizontal buoyancy, Mercker and Wiedemann used an approach based on Glauert's results (see Chapter 4.1.2) and on experimental results from actual vehicles. In the first correction version, they proposed a Glauert factor of G = 1.75 [4][5].

The first version of the complete correction method by Mercker and Wiedemann was also published in the SAE-SP by Lindener [45], including additional benchmark data. This method became the basis for further developments in open-jet correction in the following years.

In 1999, Gürtler applied the method to the database created with the EADE Correlation Test 1999 [46]. The correction results led to Gürtler's proposal to use different Glauert factors for different vehicle shapes. Further measurements in several wind tunnels hinted to a distortion of the vehicle's near field wake by different static pressure distributions, and thereby, changed measured forces. Based on measurements with the SAE model family in the IVK model-scale tunnel, Gürtler came up with a reverse calculation of the Glauert factor from measurements performed in different static pressure distributions [47]. This is possible as the correction for horizontal buoyancy is the only correction component affected by the static pressure distribution.

In 2001, Wickern proposed an alternative calculation of horizontal buoyancy of models with a bluff rear end. In his approach, the volume of the separation bubble was added to the model volume, and the effective volume was calculated with a formula proposed by Evans, which led to a Glauert factor of approximately G = 1.14 [6][38].

$$\text{Evans' Glauert factor: } G = 1 + 0.4 \frac{t}{l_M} \qquad t = 2\sqrt{\frac{2A_M}{\pi}} \qquad \text{Eq. 4.3}$$

G :   Glauert Factor

l_M :   Model length

A_M :   Model area frontal

t :   Thickness vehicle model

The volume of the separation bubble was proposed to be calculated as the volume of a paraboloid with the vehicle separation area as base area and a variable wake length. This wake length was calculated using an empirical factor, which was geometry dependent. Wickern also proposed the use of the classical correction approach for open-jets, in contrast to the semi-empirical approach by Mercker and Wiedemann [44]. In 2004, Wickern and Schwartekopp proposed a model for the nozzle interaction based on classical theory [9]. The paper included proposals for both q- and gradient correction [9][21].

In 2005, Mercker et al. proposed an approach based on Wickern's model to add the volume of the separation bubble and Gürtler's reverse calculation method, using two measurements in different pressure gradients [10]. This method later became the so-called two-step approach as the correction for horizontal buoyancy effects was separated into two parts. In the first step, classic horizontal buoyancy for the model volume was calculated including the Glauert factor, which was calculated with Evans' formula. In the second step, the horizontal buoyancy correction for the separation bubble was calculated as the pressure difference between vehicle base and the closing point of the wake bubble. While the actual wake length was geometry dependent in Wickern's approach, it became a variable in the new approach. This variable was calculated like Gürtler's Glauert factors from measurements taken in at least two different static pressure distributions. Wickern's approach with a paraboloid form of the vehicle's wake was simplified by using a cuboid.

Furthermore, Mercker et al. proposed to modify the model source location and to move the application points for the interference velocities (see Chapter 4.2.4). The model source location is the x-position of the source which generates a semi-circular body of revolution with the same base area as the frontal area of the model. This body of revolution is used to represent the vehicle, and thereby, model the deceleration of the stream into the nozzle [10]. By moving the source location in x-direction, a match between corrected measurements done with plenum and nozzle method can be achieved (see Chapter 4.2.3).

During the ongoing discussions concerning this topic, Mercker proposed the so-called one-step method that further simplified the horizontal buoyancy correction. This approach completely skipped the first part of the horizontal buoyancy correction, which was based on classic buoyancy (see Eq. 4.1). The second part was then used for the whole vehicle. This approach was published in 2006 [11], and is still used today.

The current correction method requires the simultaneous measurement of the dynamic pressure using both nozzle method and plenum method (see Chapter 2.1.2) during the measurement of the aerodynamic forces. They are necessary to calculate the singularity location x_s required for the velocity correction (see Chapter 4.2.3). Furthermore, aerodynamic forces must be measured with two different static pressure distributions present in the test section. These two static pressure distributions must be known from measurements in the empty wind tunnel test section. The method is based on the assumption that a vehicle has a typical sensitivity to the influence of the empty-tunnel pressure gradient. The sensitivity can be identified by introducing a second gradient in the wind tunnel [11].

4.2 Dynamic Pressure Correction

Four of the five interference effects mentioned above (see Chapter 3.1.) are the so-called blockage effects. These are nozzle interference, collector interference, jet deflection, and jet expansion. It is thought that these four blockage effects induce longitudinal velocity increments. Therefore, the correction of the four blockage effects is made by a correction of the velocity, or rather dynamic pressure. The jet expansion term is the classical, infinitely-long, open-jet correction. The nozzle and collector interference terms account for the finite length of the jet upstream and downstream of the vehicle [4].

All blockage effects (see Chapter 3.1) are handled by a correction of the longitudinal velocity. The velocity corrections are computed as velocity increments induced by the constraining boundaries. They have the following form:

$$U_{corr} = U_\infty + \sum_i \Delta u_i = U_\infty \left(1 + \sum_i \varepsilon_i\right) \quad \Delta u_i = U_\infty \cdot \varepsilon_i \qquad \text{Eq. 4.4}$$

U_∞ :   measured velocity

U_{corr} :   corrected velocity

Δu_i :   velocity increments for different blockage effects

ε_i :   blockage correction factor

$$\text{Basic definition of } q\text{: } q = \frac{\delta}{2} U_\infty^2 \qquad \text{Eq. 4.5}$$

$$\Rightarrow \frac{q_{corr}}{q_\infty} = \left(\frac{U_{corr}}{U_\infty}\right)^2 = \left[1 + \sum_i \varepsilon_i\right]^2 \qquad \text{Eq. 4.6}$$

The correction of the dynamic pressure in its actual form consists of perturbation velocities originating from blockage corrections for solid blockage, nozzle blockage, and collector blockage (see Eq. 4.7). The dynamic pressure correction is, of course, dependent on the method used for q determination, which is either nozzle method or plenum method.

$$\frac{q_{corr}}{q_\infty} = \left[1 + \varepsilon_S + \varepsilon_C + \varepsilon_{N/P}\right]^2 \qquad \text{Eq. 4.7}$$

ε_S :   Solid-blockage correction factor

ε_C :   Collector effect correction factor

$\varepsilon_{N/P}$:   Nozzle effect correction factor (N: nozzle method, P: plenum method)

The calculation of the different blockage correction factors is described in the following paragraphs.

4.2.1   Solid Blockage

The solid-blockage term corrects the blockage effects of jet expansion and jet deflection (see Chapter 3.1). It is shown in by formula Eq. 4.8.

$$\varepsilon_S = \frac{\tau \cdot A_M \sqrt{\dfrac{V_M}{l_M}}}{\left(\dfrac{A_N}{1 + \varepsilon_{QN}}\right)^{3/2}} \qquad \text{Eq. 4.8}$$

τ :   Tunnel shape factor

ε_{QN} :   Nozzle blockage effect correction factor

The tunnel shape factor τ is the standard open-jet blockage constant that can be calculated by a double numerical summation of images as described by Wüst [48]. Further information can also be found in [18][17].

The term $(1 + \varepsilon_{QN})$ represents the effect of the jet deflection near the nozzle with ε_{QN} being the perturbation velocity coming out of the potential-flow model for calculating the nozzle blockage effect [4][5]. Jet deflection leads to an additional jet expansion due to the flow confinement and the proximity of the nozzle. This effect is handled by using a reduced cross-sectional nozzle area for the solid-blockage correction [4].

4.2.2   Collector Effect

As described in Chapter 3.1.4, collector blockage is a similar effect, like nozzle blockage. It results from a flow acceleration that occurs in the collector due to the blockage effect of the vehicle's wake as it enters the collector and, thereby, constrains the surrounding flow by displacement effects. This solid-wall blockage effect is experienced by the vehicle as a far-field interference effect which must be corrected. Its magnitude can be determined by calculating the axial flow acceleration due to this effect. For this purpose, the nozzle is replaced by a vortex ring. The induced axial velocity can be calculated using Biot-Savart principles [4]. In potential-flow theory, the Biot-Savart law is used to calculate the velocity induced by vortex lines.

In the original work of Mercker and Wiedemann [4], it seemed to be justified to adopt the wake correction formula Eq. 4.9 for closed-wall wind tunnels as derived by Mercker in 1986 [41]. He suggested a correction formula based on the classical approach of Glauert [49] and Maskell [50]. In formula Eq. 4.9, the first term in the brackets describes the blockage effects of the far field wake (ffw), whereas the second term accounts for the near field wake using an empirical near field wake constant of $f_{nfw} = 0.41$, which was determined by Mercker [41][37].

$$\varepsilon_W = \frac{A_M}{A_C}\left(\frac{CD_{meas}}{4} + f_{nfw}\right) \qquad f_{nfw} = 0.41 \qquad \text{Eq. 4.9}$$

ε_W :   wake blockage effect correction factor

f_{nfw} :   near field wake constant

l_{TS} :   Wind tunnel test section length

CD_{meas} :   measured drag coefficient

In 1996, Mercker and Wiedemann [4] had already discussed the relevance of the adopted near field wake constant in open-jet test sections and argued that it might not be relevant if the near field wake did not expand into the collector. This could be the case for automotive wind tunnels with long test sections or fast back vehicles, which often have a considerably smaller near field wake, as compared to a notch back or square back type of car [4]. In these cases, Mercker and Wiedemann suggested to neglect this term:

$$f_{nfw} = 0 \Rightarrow \varepsilon_W = \frac{A_M}{A_C}\left(\frac{CD_{meas}}{4}\right)$$

Eq. 4.10

The near field wake constant of $f_{nfw} = 0.41$ has been part of the correction method for all vehicle rear end shapes. This is debatable even more today, as newly built open-jet wind tunnels usually feature very long test sections [51]. Furthermore, corrections of drag measurements done with formula Eq. 4.9 will lead to lower corrected coefficients than corrections using formula Eq. 4.10.

The potential-flow outside the wake is accelerated by the flow-displacement caused by the wake. This speed increase is seen by the model [4],[11] and the corresponding correction factor can be determined by applying Biot-Savart principles [4].

$$\varepsilon_C = \frac{\varepsilon_W \cdot r_C^3}{\left(r_C^2 + \left(l_{TS} - x_M - \frac{l_M}{2}\right)^2\right)^{3/2}} \qquad r_C = \sqrt{\frac{2A_C}{\pi}}$$

Eq. 4.11

ε_W :   Wake blockage effect correction factor

r_C :   Equivalent radius of the duplex collector area

l_{TS} :   Wind tunnel test section length

x_M :   Model x-position (distance of wheelbase center to nozzle exit plane)

4.2.3   Nozzle Effect

The proximity of the nozzle to the test object causes the effective dynamic pressure at the model location to be increased above the measured (virtual) approach velocity (see Chapter 3.1.2) [11]. The magnitude of this effect is different for the nozzle method and the plenum method, which have been described in Chapter 2.1.2. The interference is different for both measurements in the presence of a model near the nozzle, with neither giving the true nozzle speed. When the vehicle is far from the nozzle, both methods return identical dynamic pressure measurements. When a vehicle is near the nozzle, the two methods diverge. The latest version of the correction procedure requires both nozzle and plenum dynamic measurements to be made. Then, both measurement results are corrected to the same value [11].

Mercker and Wiedemann identified the nozzle effects as a solid-wall blockage effect, which is comparable to the blockage situation in a closed test section and experienced by the model as a far-field interference effect [4][5]. Therefore, the induced velocity at the model position can be calculated using Biot-Savart principles by replacing the nozzle with a vortex ring. The result can be seen in formulas Eq. 4.12 and Eq. 4.13.

$$\varepsilon_N = \frac{\varepsilon_{QN} \cdot r_N^3}{\left(r_N^2 + \left(x_M - \frac{l_M}{2} \right)^2 \right)^{3/2}} \qquad \text{Eq. 4.12}$$

$$\varepsilon_P = \frac{\varepsilon_{QP} \cdot r_N^3}{\left(r_N^2 + \left(x_M - \frac{l_M}{2} \right)^2 \right)^{3/2}}$$

Eq. 4.13

ε_N :   Nozzle blockage effect correction factor (Nozzle method)

ε_P :   Nozzle blockage effect correction factor (Plenum method)

r_N :   Equivalent radius of the duplex nozzle area

x_S :   x-position of the source that is used to calculate the interference effect

The blockage effect inside the nozzle is represented by ε_{QN} and ε_{QP}, respectively (see formulas Eq. 4.14 and Eq. 4.15). This has been determined by Mercker and Wiedemann using a simple potential-flow model that consists of a point source exposed to a parallel flow. The result is a half-infinite body of revolution which represents the test vehicle plus its mirror image [4][5].

$$\varepsilon_{QN} = \frac{\dfrac{A_M}{2A_N}\left(1 - \dfrac{x_S}{\sqrt{x_S^2 + r_N^2}} \right)}{1 - \dfrac{A_M}{2A_N}\left(1 - \dfrac{x_S}{\sqrt{x_S^2 + r_N^2}} \right)}$$

Eq. 4.14

$$\varepsilon_{QP} = \frac{\dfrac{A_M}{2\pi}\left(\dfrac{x_S}{\sqrt{x_S^2 + r_N^2}^{\,3}}\right)}{1 - \dfrac{A_M}{2\pi}\left(\dfrac{x_{S,1/2}}{\sqrt{x_S^2 + r_N^2}^{\,3}}\right)}$$

Eq. 4.15

$$r_N = \sqrt{\frac{2A_N}{\pi}}$$

Eq. 4.16

The first model of the flow distortion in the nozzle, by Mercker and Wiedemann [5], did not produce fully-satisfactory results [10]. After the correction was applied, the nozzle method and the plenum method should have produced the same dynamic pressure, which was not necessarily the case with the first model. In 2005, Mercker et al. [10] solved this problem by adjusting the position x_S of the source that was used to estimate the nozzle blockage. The arbitrary choice of setting the stagnation point of the semi-infinite body at the front bumper location of the model was changed to a recalculation of x_S from measurements using both nozzle method and plenum method. The source position x_S is treated as a variable and calculated to fulfill equation Eq. 4.17.

$$\left(1 + \varepsilon_P\right)^2 \cdot q_P = \left(1 + \varepsilon_N\right)^2 \cdot q_N$$

Eq. 4.17

4.2.4   Application Points for the Interference Velocities

The Mercker-Wiedemann method [4][5] proposed to apply the interference velocities from both the nozzle and collector interference corrections at the model center. This choice was arbitrary, but based on the aeronautical approach where the model is usually small compared to the test section and has a small wake [10]. It is obvious, that not only the blockage ratio of typical automobile models in typical wind tunnel test sections is different from the aeronautical approach, but also the ratio of model length and test section length is much bigger. As mentioned in Chapter 4.1, in 2005, Mercker et al. [10]

proposed to move these application points for nozzle and collector blockage correction to the vehicle's front and rear bumper location, respectively. This choice was also arbitrary and based on empirical findings. These revised positions are still used today.

The choice of these application points will not affect the correction method's ability to drastically reduce the spread of measured coefficients obtained in one wind tunnel, for example, with different static pressure distributions or different methods for q determination. But, it will definitely affect the final level of the corrected coefficients. The closer the application points are set to nozzle and collector, respectively, the bigger the blockage correction factors will become and, therefore, the smaller the corrected coefficient will be calculated. Measurements from several different wind tunnels with different q determination methods, and for different static pressure distributions, could help determine the final position. Unfortunately, a complete database of such measurements does not exist, to the best of this author's knowledge.

4.3 Horizontal Buoyancy Correction

The fifth interference effect is so-called horizontal buoyancy which has been described in Chapter 3.2. The historical development of the correction of this effect has been described in Chapter 4.1. As mentioned, Mercker simplified Wickern's approach of a paraboloid form as the vehicle's shape, by using a cuboid. In 2006 [11], it was proposed to use this cuboid as a representation of the complete vehicle, including the near field wake. The frontal area of the cuboid is identical with the vehicle's frontal area. Its length is dependent on the variable x_{WE}, which represents the position of near field wake end. This position is not a distinct position anymore, which can be found by using several different methods, as proposed in a previous paper by Mercker et al. [10]. It is a variable and is recalculated using two measurements in two different static pressure distributions in one wind tunnel. This means, that x_{WE} is varied until the final corrected drag values from two different static pressure distributions are equal. Therefore, it is assumed, that x_{WE} is the same for each vehicle in different static pressure distributions.

The final correction term for horizontal buoyancy is given by

$$\Delta CD_{HB} = cp_{stat}(x_{WE}) - cp_{stat}(x_{MF})$$ Eq. 4.18

ΔCD_{HB} :   Correction term for horizontal buoyancy

$cp_{stat}(x)$:   Static pressure distribution at x-position

x_{WE} :   x-position at end of vehicle wake

x_{MF} :   x-position at end of vehicle model front

4.4   Open-jet correction method

The combination of both correction components for blockage effects and horizontal buoyancy leads to the following formula Eq. 4.19, which represents the complete correction formula for open-jet interference effects.

$$CD_{corr} = \frac{CD_{meas} + \Delta CD_{HB}}{\dfrac{q_{corr.}}{q_{\infty}}}$$ Eq. 4.19

The correction terms for both components are calculated using formulas Eq. 4.7 and Eq. 4.18.

The actual application of the method is separated into two sequential steps. In the first step, the singularity location x_S is calculated. In the second step, the effect of the static pressure distribution is corrected. A step-by-step description of the application of the open-jet correction method is located in Appendix A.3.

5 Computational Fluid Dynamics Investigations

In this thesis, CFD simulations were used to investigate the blockage effects in an open-jet test section, which significantly change the flow field and, thus, the acting forces on the vehicle. Hence, it is the goal to gain deeper insight into single interference effects and, finally, to gain deeper understanding into the physics behind the empirical correction method for open-jet interference effects. The main focus is on the horizontal buoyancy effect, which is the strongest of all open-jet interference effects in most known open-jet wind tunnels.

It was necessary to ensure that all effects were captured correctly in a simulation, including the whole wind tunnel test section with a vehicle model present. This required extensive validation with wind tunnel measurements, including forces and flow fields.

Therefore, a CFD simulation setup capable of reproducing all effects present in the IVK model-scale wind tunnel (IVK-MWK) in Stuttgart was developed. This specific wind tunnel was chosen because of various known methods to change the static pressure distribution in the test section [52].

All simulations were conducted using Exa's PowerFLOW (see Chapter 2.2), which was selected because of its transient nature, due to the underlying Lattice-Boltzmann algorithm. Therefore, it had the capability to capture all turbulence effects in the highly turbulent shear layer and the flow field around the model.

5.1 Basic CFD Investigation of Horizontal Buoyancy

The interference effect of horizontal buoyancy is the most dominant of all open-jet interference effects in the majority of automotive wind tunnels. Therefore, a generic case was simulated, first, to investigate the effect of a

© Springer Fachmedien Wiesbaden GmbH, part of Springer Nature 2018
O. Fischer, *Investigation of Correction Methods for Interference Effects in Open-Jet Wind Tunnels*, Wissenschaftliche Reihe Fahrzeugtechnik Universität Stuttgart, https://doi.org/10.1007/978-3-658-21379-4_5

static pressure distribution on a test body in a simple setup which can be seen in Figure .

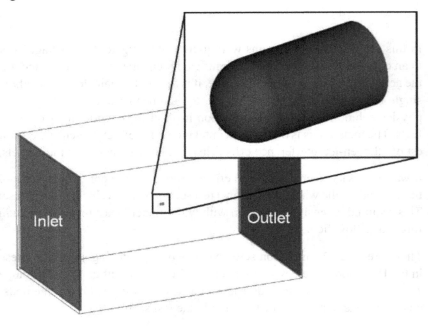

Figure 5.1:   Simple computational fluid dynamics simulation to investigate horizontal buoyancy (closed-wall wind tunnel and test body)

A closed-wall wind tunnel setup was chosen for this investigation to guarantee defined boundary conditions. It will be referred to as "box." It has a symmetric rectangular extension, relative to the body and perpendicular to flow direction. All walls were set to boundary condition "frictionless" to prevent any boundary layer development along the walls. To prevent any closed-wall blockage effects from influencing the flow field, a low blockage ratio (see equation Eq. 3.1) of 0.022% was chosen. The lift and side forces were zero for all cases, due to the symmetric wind tunnel geometry.

Two different test bodies were used: (a) a bluff body with flow separation at the base area, and (b) a body with very small separation at the end, due to a streamlined extension (see Figure 5.2). Both bodies consisted of a sphere and a cylinder; however, the streamlined body had a body extension shaped like the separation bubble of the simple body.

Figure 5.2: Bluff (left) and streamlined (right) test bodies

The static pressure distribution in the box without a test object present is constant over the whole test section. Therefore, different static pressure distributions had to be generated which was done by using a spherical stagnation body downstream of the test body position. A variety of static pressure distributions acting on the test body could be generated by changing the size of the stagnation body (radius of sphere) and by moving the test object in x-direction through the non-linear distribution (see Figure 5.3).

The different static pressure distributions could be sorted into three groups. The first one included small static pressure gradients like Numbers 1, 2, and 3. The second group included static pressure gradients, which are representative for static pressure distributions in typical automotive wind tunnels (Numbers 4 and 5). The third set included an example for a strong static pressure distribution (Number 6).

The results for drag were absolutely consistent with the expected effect of the static pressure distribution on forces (see Figure 5.4). Furthermore, the correction for horizontal buoyancy, as described in Chapter 4.3, was applied to the simulation results (see Figure 5.4).

A comparison of the drag coefficients for the two different body shapes shows the expected result. The streamlined body had a much lower drag, as compared to the bluff body (see Figure 5.4). The drag was reduced by the different static pressure distributions for both body shapes. The static pressure gradient Number 0 denotes the case with a constant static pressure distribution in the test section and, therefore, has no horizontal buoyancy effect. The static pressure gradients with Numbers 1, 2, and 3 show small effects on both test bodies, while the gradients 4 and 5 have significant effects. The static pressure gradient Number 6 has very strong effects on the measured drag.

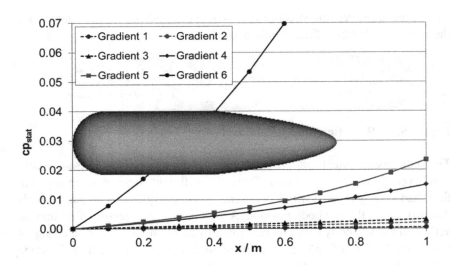

Figure 5.3:   Static pressure distributions for the basic computational fluid
                  dynamics investigation of horizontal buoyancy

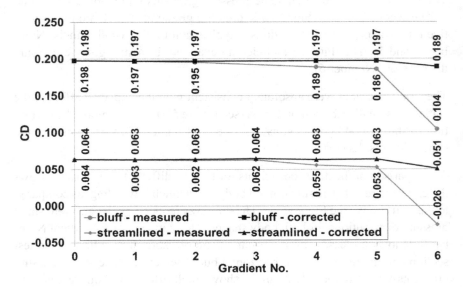

Figure 5.4:   Measured and corrected drag coefficients for the two test bod-
                  ies simulated in different static pressure distributions

The correction has been made using the horizontal buoyancy described in Chapter 4.3. All effects of the static pressure gradients up to Number 5 can be perfectly corrected with the correction method. The correction also shows the right trend for gradient Number 6, however, is not capable of fully correcting the measured value. This might be caused by uncorrectable effects in the flow field around the test body, due to the strong pressure gradient.

This basic CFD investigation revealed that horizontal buoyancy can be simulated with PowerFLOW. Furthermore, the correction component for horizontal buoyancy does work for the simulated static pressure distributions. As these pressure distributions can be representative of typical automotive wind tunnels, the simulation of a full model of an open-jet wind tunnel, including the effects of horizontal buoyancy, is possible.

For the simulation of an open-jet wind tunnel, it is essential to match the wind tunnel geometry in order to produce the same effects as in the real wind tunnel; therefore, it has been recognized that a very detailed reproduction of the IVK-MWK was necessary to ensure that all effects could be simulated correctly for different vehicles.

5.2 Validation Measurements in IVK-MWK

A variety of validation measurements for the CFD simulations were carried out in the IVK-MWK (see also Appendix A.2.1) in order to validate simulation with and without a vehicle model present in the test section.

The tunnel was equipped with a five-belt rolling road system, including several systems for boundary layer control, such as suction and tangential blowing [53]. Since the simulation of the effects of such installations is not part of the current thesis, all experiments and simulations were performed without any ground simulation. Suction areas in the test section floor, including a boundary layer presuction (BLPS) and a distributed boundary layer suction (DBLS), were taped to prevent any volume flow through the inactive systems. Furthermore, gaps between center belt (CB) and the surrounding test section floor at front and rear roller were also taped, including the tangential blowing in front of the center belt (TBCB). This is necessary, as these gaps may influence the static pressure distribution in lower measuring heights close to the test section floor (see Figure 5.5 and Figure 5.6). This influence becomes negligible for larger measuring heights (z-figures, see Figure 5.7). These changes would not be captured by the simulation, as the gaps have not been modeled in the digital wind tunnel geometry.

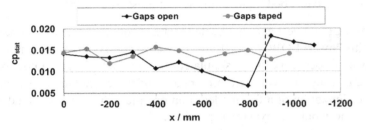

Figure 5.5: Measured static pressure distribution (z = 50 mm), rear center belt roller gaps at dotted line

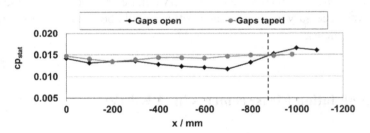

Figure 5.6: Measured static pressure distribution (z = 150 mm), rear center belt roller gaps at dotted line

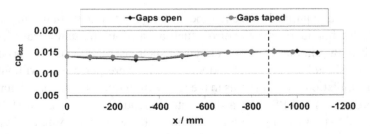

Figure 5.7: Measured static pressure distribution (z = 450 mm), rear center belt roller gaps at dotted line

All boundary layer measurements were performed with a special pitot probe. Further flow field measurements were conducted with various techniques. For several wake measurements at the SAE model, a pitot rake was used. Flow fields, themselves, were also investigated with a fast response multi-hole pressure probe known as "Cobra Probe" [54]. These measurements help to understand flow velocities and flow angles in the flow field. Several measuring planes were placed around the models and in the near wake. The grid for all measuring planes was a 10 mm square. Due to the fast response characteristics of the pressure probe, the so-called "Flying Probe" technique was used to determine the velocity components of the flow field [55]. As a consequence of the limited acceptance cone of the probe (± 45°), the recirculation flow inside the near field wake could not be captured. Therefore, the near field wake area is marked as a checkered or striped area on the plots.

All forces and moments acting on the vehicle models were measured with an external six-component pyramidal balance, located underneath the test section floor [56].

5.3 Simulation of IVK-MWK (DIVK)

The flow conditions in the wind tunnel test section without a vehicle model present had to be validated by various comparisons between simulation and experiment. This was required to ensure that all interference effects were simulated correctly when a model was placed into the test section.

5.3.1 Digital Wind Tunnel Model (DIVK)

A detailed digital reproduction of the IVK-MWK was created, including a plenum chamber with test section, nozzle, collector, and traversing system (Figure 5.8) [1][57]. A return duct and turning vanes are not part of the digital model of IVK-MVK (DIVK). All dimensions of the virtual wind tunnel correspond to the test section dimensions of the actual facility.

The influence of the contraction on the flow around the vehicle model has been investigated in several simulations. The simulation results of this wind tunnel

(IVK-MWK) showed, that the simulated contraction generates negligible influence on the flow in the test section. Therefore, it was not used for the simulations with vehicle models, in order to reduce simulation times. The contraction was replaced by a straight duct. Generally speaking, the simulation of the contraction geometry might be necessary in other wind tunnel geometries to reproduce all effects.

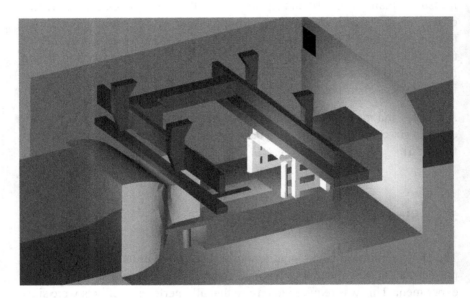

Figure 5.8:   Digital model of IVK model-scale wind tunnel with nozzle contraction (left) and diffuser (right)

Turbulence generators close to the nozzle exit plane have a strong influence on the shear layer development and the static pressure distribution along the wind tunnel test section. For the IVK-MWK, the influence of several different turbulence generator configurations on static pressure distribution along the wind tunnel test section can be seen in Figure 5.9. These configurations differ in geometry (Seiferth-wings and delta-wings, see Figure 5.10) and x-position of the turbulence generators at the nozzle walls. The big differences in static pressure distribution emphasize the necessity of a correct shear layer development.

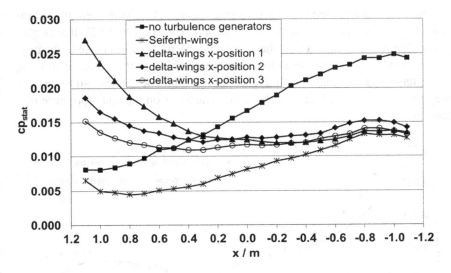

Figure 5.9: Influence of different turbulence generator configurations in IVK-MWK

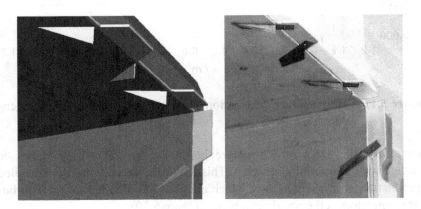

Figure 5.10: Delta-wings (simulation model [left], real geometry [right]) at nozzle exit plane

Several investigations were carried out with various turbulence generator geometries to investigate the effects on flow structure of the shear layer and static pressure distribution. The static pressure distribution could be matched using simple cubes as turbulence generators (see Figure 5.11). They were placed inside the nozzle, close to the nozzle exit plane (see Figure 5.12). It was also found that it is possible to tune the static pressure distribution close to the nozzle by moving the boxes in x-direction (comparable to the effect of moving the delta-wings, see Figure 5.9). Hence, using turbulence boxes can be an easy way to simulate the effect of turbulence generators on the static pressure distribution.

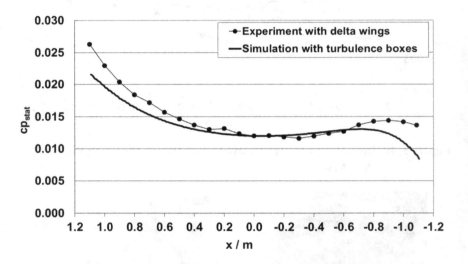

Figure 5.11: Static pressure distribution using wing boxes as turbulence generators in DIVK

Conversely, the exact flow structure inside the shear layer could not be matched with this simulation setup. This led to the decision to use a detailed reproduction of the delta-wings, which are used in the IVK-MWK, as turbulence generators at the nozzle exit plane (Figure 5.10).

Several additional simplifications of the digital wind tunnel model could be applied after investigations on the influence of traversing system and plenum ventilation. The traversing system was abandoned, as it had no influence on the flow inside and close to the open-jet in this wind tunnel. This is probably because the plenum of the IVK-MWK is relatively large, as compared to the open-jet. This might not be true for other wind tunnels, especially full-scale tunnels with smaller plenum sizes. Furthermore, additional ventilation openings were added to reduce the computation time for the initial stabilization of the flow, until a constant pressure was established in the plenum (see Figure 5.13). It was crucial to ensure that the jet was not influenced by the additional ventilation openings. The ventilation of the plenum in the real IVK-MWK is done through a small opening in the upper left corner of the plenum wall above the collector. This opening is also represented in DIVK. The computational domain size of the simulation is limited by the virtual wind tunnel model. Both nozzle and diffuser ducts were extended in flow direction to prevent any influence on flow inlet and outlet.

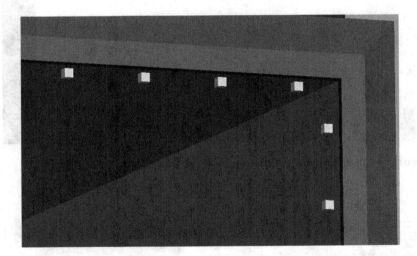

Figure 5.12: Simple cubes were tested to be used as turbulence generators in DIVK

The computational mesh of the simulation is refined in shear layer areas and boundary layer regions to resolve the flow structures accurately (Figure 5.14 and Figure 5.15). The smallest voxel in the shear layer is 3 mm for all cases. The simulation of the standard wind tunnel, without vehicle model, consists of 67 million voxels. The simulation, including the stagnation bodies, consists of 73 million voxels.

Figure 5.13: Additional ventilation inlets in DIVK (white)

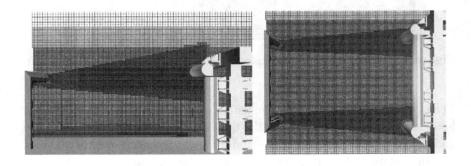

Figure 5.14: Computational mesh of test section without vehicle (left y = 0, right z = 0.45 m): high resolution for shear layer development and boundary layer

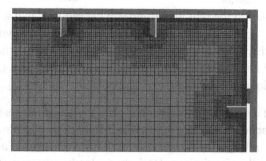

Figure 5.15: Computational mesh around delta-wings at nozzle exit

5.3.2 Boundary conditions

The right choice of boundary conditions in a CFD simulation may influence the results significantly. In the geometry used for the simulation of IVK-MWK, three major boundary conditions must be chosen (apart from walls with different roughness and friction):

■ plenum ventilation,

■ inlet upstream of the nozzle, and

■ outlet downstream of the collector.

The plenum ventilation in a real wind tunnel connects the plenum with the atmosphere surrounding the wind tunnel building. It is designed to maintain a constant pressure inside the plenum in all wind tunnel operating conditions. Therefore, the plenum ventilation has to be established with the boundary condition "Static Pressure."

As described in Chapter 2.1.2, the correct determination of wind velocity and, thereby, reference dynamic pressure is of fundamental importance, as it is necessary to derive dimensionless quantities from wind tunnel force measurements. Also, it is fundamental that a comparison of measurement and simulation results have equal wind speeds.

The easiest way to determine the wind velocity in the simulation would be to use either plenum method or nozzle method (see Chapter 2.1.2), according to the procedure used in the real wind tunnel. This requires that a settling chamber and a nozzle contraction be part of the wind tunnel geometry used in the simulation. Furthermore, boundary layers developing along the nozzle walls have to be simulated, which requires sufficient resolution and, therefore, additional computational effort.

The influence of a nozzle contraction has been investigated. As mentioned, the simulated contraction generated negligible influence on the simulation results. Based on these findings, the nozzle contraction was dispensed and replaced by a straight duct in the simulations. Both ducts, leading from nozzle to inlet, as well as from collector to outlet, respectively, have been extended to prevent any direct influence of the boundary conditions on the flow field.

As the contraction was not simulated for all cases, both inlet and outlet have been chosen to be setup with the boundary condition "Mass Flow." As mentioned in Chapter 2.1.2, the nozzle method produces a constant volume flow through the nozzle exit plane, as long as the adverse pressure gradient in front of the model does not affect the measured nozzle pressure. The mass flow has been calculated from the chosen test speed of 50 m/s. The exact calculation procedure can be found in Appendix A.4. The chosen boundary conditions for the DIVK simulation are shown in Table 5..

Table 5.1:   Boundary conditions in DIVK

Face	Boundary condition
Inlet (upstream of the nozzle)	Mass flow
Outlet (downstream of the collector)	Mass flow
Plenum ventilation	Static pressure (101,300 Pa)
Walls, floor	Standard wall (roughness 0)
Center belt	Standard wall (roughness 0.05 mm)
Nozzle walls close to nozzle exit plane	Standard wall (roughness 1 mm)
Duct extensions upstream of the nozzle and downstream of the collector	Frictionless wall

Turbulence intensity and turbulence length scale were set to 0.003 mm and 1.5 mm, according to the free stream values of the wind tunnel.

5.3.3 Boundary Layer

The correct reproduction of the boundary layer in the simulation is important for the prediction of both drag and lift. Comparative aerodynamic force measurements of production cars, with and without boundary layer control, have shown that the influence of boundary layer variations is even stronger on lift forces than on drag forces [37].

The IVK-MWK is equipped with a five-belt system for road simulation which was not in operation for the experimental data shown here and, therefore, not simulated in CFD. But, for the correct simulation of the boundary layer development along the test section, the roughness of the inactive CB had to be taken into account by defining an area with higher roughness, as compared to standard wall conditions on the steel test section floor surrounding the CB. The necessity of the correct roughness has been shown by Kuthada et. al. [58].

Several numerical investigations were carried out to match the boundary layer development along the test section with the measured profiles [1]. The simulated boundary layer profiles in the empty test section match the measured profiles in the vehicle's vicinity (625 mm > x > -625 mm) as can be seen in Figure 5.16 and Figure 5.17. The agreement between measured and simulated boundary layer development can be considered sufficient. The remark, "configuration 1," refers to the static pressure distribution, as explained further in Chapter 5.3.5.

A noticeable influence on boundary layer development was not observed when the modified static pressure distribution of configuration 2 or configuration 3 was attained with the help of stagnation bodies in the collector. Therefore, only the result of configuration 1 is shown here.

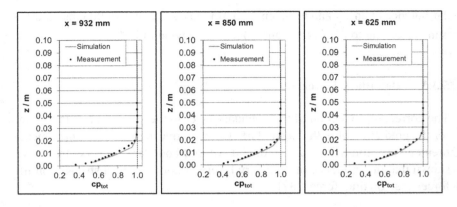

Figure 5.16: Comparison of boundary layer in test section simulation and measurement (configuration 1)

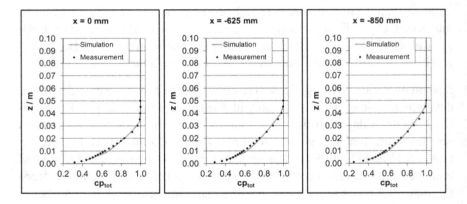

Figure 5.17: Comparison of boundary layer in test section simulation and measurement (configuration 1)

5.3.4   Shear Layer

The static pressure distribution along the test section is strongly influenced by the shear layer development. Therefore, a detailed representation of the above mentioned delta-wings is used to simulate the correct shear layer development

[1]. The computational mesh around the detailed delta-wings is shown in Figure 5.15. Figure 5.18 shows the shear development along the test section, including two planes at two x-positions, where experimental data are available.

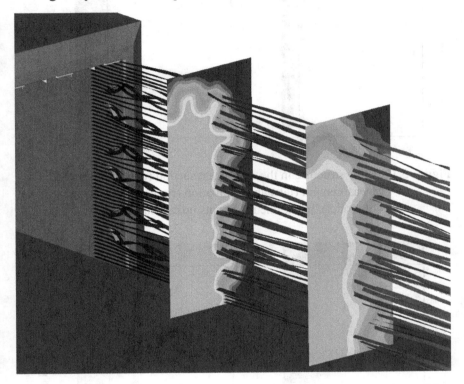

Figure 5.18: Simulation of shear layer development along test section, with detailed delta-wings (configuration 1), displayed planes at x = ±0.5 m

A comparison of the resulting flow structure of the shear layer at the two x-positions shown in Figure 5.18 is displayed in Figure 5.19 and Figure 5.20. Experimental data and simulation results refer to the standard wind tunnel configuration without stagnation bodies.

The figures show dimensionless x-velocities, which were measured with a pitot static probe in y-z-planes in the real wind tunnel. The added black line in all plots represents the nozzle size.

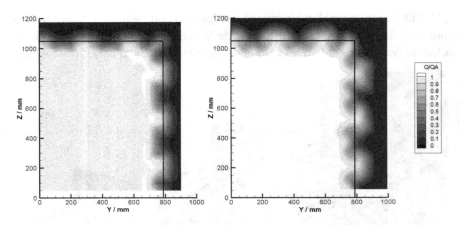

Figure 5.19: Flow structure in the empty test section (configuration 1), measurement (left) and simulation (right), $x = 0.5$ m (near front of model), black line represents nozzle size

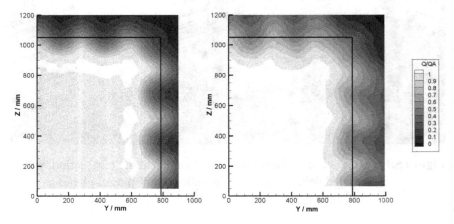

Figure 5.20: Flow structure in the empty test section (config 1), measurement (left) and simulation (right), $x = -0.5$ m (near rear end of model), black line represents nozzle size

The comparison shows an agreement of the basic flow structure of experimental and numerical results, but still some discrepancies in the detail. Similar results have been seen in measurements done with the Cobra Probe [54]. The simulated shear layer also seems to grow faster than the measured one, but only as relating to the outer boundary (Figure 5.20). The inner boundary, which is important for the development of static pressure distribution, shows an acceptable agreement between simulation and experiment.

5.3.5 Static Pressure Distribution

The static pressure distribution of the standard wind tunnel configuration in the IVK-MWK is not a typical distribution for an automotive wind tunnel. The static pressure is nearly constant over the whole test section length apart from a small acceleration of the flow discharged by the nozzle. Therefore, horizontal buoyancy effects are small in the IVK-MWK, in its standard configuration. It is even possible to avoid this acceleration with a further improved configuration of today, reaching a completely constant static pressure distribution.

Several methods for changing the static pressure distribution in the test section of the IVK-MWK can be used to investigate horizontal buoyancy effects. One such method was used to represent static pressure distributions (which are more typical for an automotive wind tunnel), meaning there was a significant rise of static pressure close to the collector. This was achieved by using so-called stagnation bodies, which are half-round, wooden bodies (see Figure 5.21). They can be mounted to the collector walls at two different stream-wise (x-axis) positions and be setup in several heights. They basically reduce the effective collector cross-section. Depending on the chosen height of the stagnation bodies, the cross-section can be reduced anywhere from 7.09% to 16.09%. Furthermore, a strong interaction of stagnation bodies and nearby collector breathers could be found during the investigations. Therefore, it was decided to close the collector breathers for all performed measurements.

Overall, three different configurations were used throughout all measurements. The wind tunnel configuration, without stagnation bodies and with taped breathers, will be referred to as configuration 1. Two configurations, generated with stagnation bodies, will be referred to as configuration 2 and configuration 3 (Figure 5.22). They have been generated with two different heights of the stagnation bodies, mounted at the same position. Detailed information about the configurations used can be found in Appendix A.2.1.

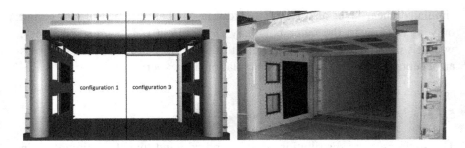

Figure 5.21: IVK-MWK collector: left simulation (with/without stagnation
bodies mounted), right experiment with stagnation bodies
mounted

There are four breather openings in each of the collector walls (see Figure
5.21). They have been closed in all configurations to prevent strong interaction
with the stagnation bodies, which are mounted nearby. The closing has no effect in the wind tunnel standard configuration (configuration 1) on the static
pressure distribution.

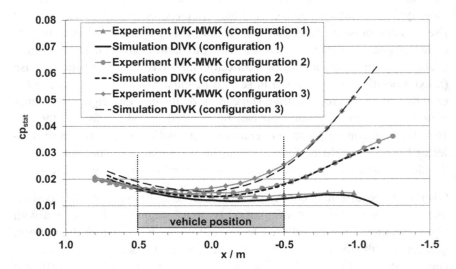

Figure 5.22: Comparison of static pressure distributions generated with stagnation bodies

The comparison of measured and simulated static pressure distributions for the three wind tunnel configurations are shown in Figure 5.22. A good reproduction of the static pressure distribution was achieved across the range of the vehicle model including its near field wake for all configurations, which is important for the correct simulation of horizontal buoyancy. The delta-wings have been moved 30 mm inside the nozzle, relative to the original position, to improve the match close to the nozzle. Small differences might result from gaps and screws in the test section floor, which were not simulated, possibly influencing the measurement.

5.4 CFD Simulation of Different Vehicle Models

Several vehicle models were simulated within different simulation setups to study blockage effects in IVK-MWK. The first simulation setup was the vehicle model inside DIVK, in its standard configuration (configuration 1). The second and third cases were the vehicle model inside the IVK model-scale wind tunnel (IVK-MWK) in its configuration 2 and configuration 3. These cases were used to validate and investigate blockage effects in IVK-MWK and, especially, horizontal buoyancy effects.

The fourth simulation setup was the vehicle model in a typical CFD setup, as it used in standard vehicle development. In this setup, the boundary conditions are far enough away from the model position to have little influence on the flow around the model. PowerFLOW refers to this setup as the Digital Wind Tunnel (DWT). It is used in this investigation to get blockage-free results for each model. This makes it possible to compare blockage-free results with corrected values from the real wind tunnel. Detailed information and pictures of the models used in this investigation can be found in Appendix 9.

The resolution setup close to the models are identical in all CFD simulations, except for one model, in order to make simulation results comparable. The floor boundary condition is set to provide a development of the boundary layer to match the thickness measured in the experiment.

5.4.1   SAE Squareback

The SAE squareback model was simulated because a best practice setup was already available from simulations in DWT [59]. The SAE squareback model was tested and simulated in the wind tunnel configurations 1 and 3. Results for configuration 2 are not available for this model. Further information, including pictures of the model, can be found in Appendix A.1.

Several additional regions with finer simulation mesh were added, according to the best practice setup. The simulation with vehicle model consisted of about 235 million voxels. The simulation with standard static pressure distribution ran for approximately 21,000 CPUh, excluding discretization. The model was tested and simulated in configurations 1 and 3.

5.4.1.1   Centerline Pressure Distribution

In Figure 5.23 and Figure 5.24, the pressure distributions along the centerline (y = 0) of the model surface were compared to experimental results.

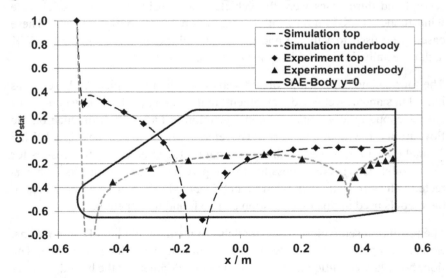

Figure 5.23: Centerline pressure distribution on SAE squareback model with standard static pressure distribution (configuration 1)

The pressures on top and underbody for configuration 1 were in good agreement with the measured values in Figure 5.23. Small differences could be observed in the model's diffuser. In Figure 5.24, bigger differences for configuration 3 were visible on the model's top side, while the pressures in the diffuser seemed to be in better agreement.

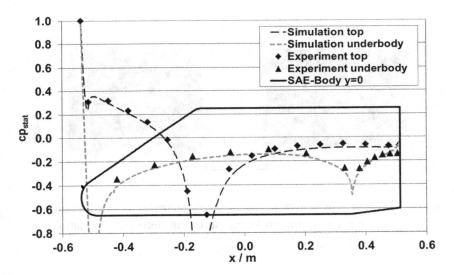

Figure 5.24: Centerline pressure distribution on SAE squareback model with standard static pressure distribution (configuration 3)

5.4.1.2 Total Pressure in Near Field Wake

In Figure 5.25 and Figure 5.26, total pressure measurements in the near field wake of the SAE model were compared to CFD results. The data used for the comparison were extracted from CFD results at the exact same positions as measured in the experiment. These were only results from configuration 1, as such data were not measured for configuration 3. Figure 5.25 shows a plane 50 mm behind the model base. The plane displayed in Figure 5.26 coincides with the model centerline (y = 0).

The simulation results were computed with a method called virtual pressure probe (VPP) [22] to account for the mentioned angular dependency of the pitot pressure rake used for the measurement. In the uncorrected simulation results, the contribution of the reverse flow components inside the near field wake would be visible which could not be captured by the pitot pressure probes. Therefore, the VPP-method was used to enable comparison between simulation and experiment.

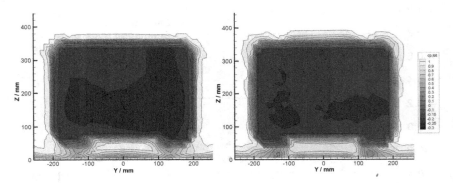

Figure 5.25: SAE squareback model: Total pressure 50 mm behind model base (configuration 1), measurement (left) and simulation [VPP] (right)

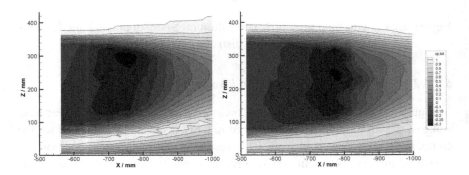

Figure 5.26: SAE squareback model: Total pressure behind model base (y = 0 mm) (configuration 1), measurement (left) and simulation [VPP] (right)

In Figure 5.25 and Figure 5.26 the measurement and simulation (VPP) results were in good general agreement. In Figure 5.25, the boundary layer development and the flow under the model seemed to match well, but a different boundary layer development on the roof was observed. In the simulation, two dips at $y = \pm 140$ mm were seen which were due to the a-pillar vortices. These vortices seemed to be smaller in the experiment. The pressure close to the model base was higher in the simulation than in the experiment (Figure 5.25) which did not correlate to the drag values (see forces analysis).

In Figure 5.26, the flow field close to the model matched between simulation and experiment. This was also the case for the lower areas ($z < 100$ m) of the flow field. In the areas above $z = 100$ m, several small differences were observed. In the measurement, the region with the lowest pressure was closer to the model base than in the simulation. This resulted in the lower pressures on the model base in the experiment as mentioned above (Figure 5.25). The widening of the wake in the upper region (starting at $x > -700$ mm) could not be seen in the simulation. The differences in Figure 5.26 at the end of the wake ($x > 800$ mm) were also the outcome of angles around 90 degrees between approach flow and the symmetry axis of the pitot pressure probe. At these angles, the pitot pressure probe was highly sensitive and, therefore, discrepancies of pressures computed with the VPP-method were most likely.

5.4.1.3   Forces

The comparison between simulated and measured integral forces of drag and lift is presented in Figure 5.27. The differences between experiment and simulation for configuration 1 amounted to 10 counts in drag, 15 counts in total lift, -7 counts in front lift, and 22 counts in rear lift. Because the simulation of the empty wind tunnel seemed to be in very good agreement with the experimental data, these differences were probably due to the vehicle setup. Possible areas for the resulting discrepancies included boundary layer development on the vehicle and the near field wake. The effects on lift balance especially need to be investigated further as the front lift was too low, and the rear lift was too high. The differences between experiment and simulation for configuration 3 amounted to 8 counts in drag, 16 counts in total lift, -10 counts in front lift, and 26 counts in rear lift. The measured and simulated drag values were lower as compared to configuration 1, as the higher static pressure difference over the vehicle caused a stronger longitudinal force which lowered the drag. This result emphasizes the importance of a correct simulation of the static pressure distribution when open-jet interference effects are simulated.

The comparison of both configurations showed that the difference in drag between simulation and experiment in configuration 3 (10 counts) was comparable to the difference in configuration 1 (8 counts). The lift balance in configuration 3 did not match the experimental data, as the front lift was too low, while the rear lift was too high. The lift deltas between experiment and simulation were comparable in both configurations. This led to the conclusion that the effects of wind tunnel interference on the acting forces were simulated correctly.

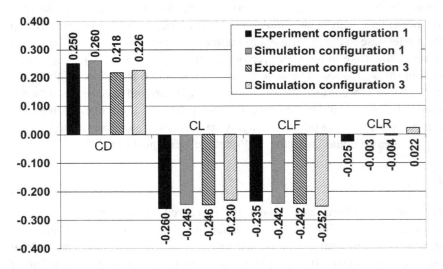

Figure 5.27: SAE squareback model: Integral values of drag and lift from simulation and experiment for both configurations (configuration 2 not available for this model)

5.4.2   Detailed Notchback (Scale 1:5)

The notchback vehicle model was a detailed scale model (1:5), built with side mirrors and a detailed underbody design, but no underhood flow. The model was tested and simulated in the wind tunnel configurations 1 and 3 [57]. Results for configuration 2 are not available for this model. Further information, including pictures of the model, can be found in Appendix A.1.2.

This detailed notchback model was designed as a CFD validation model and was equipped with 130 built-in pressure taps. The DIVK simulation with vehicle model consisted of about 205 million voxels; the DWT simulation used about 116 million voxels. The model was tested and simulated in configurations 1 and 3 (see Figure 5.22).

5.4.2.1   Centerline Pressure Distribution

Figure 5.28 and Figure 5.29 show the surface pressure measurements from experiment (configuration 3), DIVK (configuration 3) and DWT cases [57]. Figure 5.28 shows the comparison on the vehicle (top side) in the centerline of the model (y = 0 mm), while Figure 5.29 shows the pressures off-centerline (y = 60 mm). A comparison of the underbody pressures can be found in Figure and Figure A. (see Appendix 8.6). The highest difference in the surface pressure of the DIVK and experiment was a cp value of 0.08 in the highest gradient area on the hood. Overall, deviations were quite small and, in general, the DIVK results were in better agreement with the experiment, as compared to the DWT results.

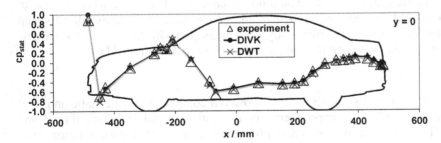

Figure 5.28: Detailed notchback (scale 1:5): Surface pressure measurements between experiment (configuration 3), DIVK (configuration 3) and DWT, at the centerline (y = 0 mm), top side of the model

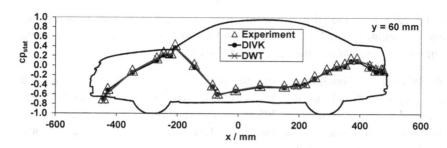

Figure 5.29: Detailed notchback (scale 1:5): Surface pressure measurements between experiment (configuration 3), DIVK (configuration 3) and DWT, at y = 60 mm, top side of the model

5.4.2.2   Flow Field

The figures in this section showed the static and total pressure in the plane "X = 500 mm," including the physical location of this plane relative to the vehicle model [57]. The simulation data were extracted at exactly the same measurement positions as the experimental data to ensure the best comparability.

As mentioned in Chapter 5.2, the Cobra Probe [54] used for these measurements had a limited acceptance cone of ± 45°. Therefore, the recirculation flow inside the near field wake could not be captured. The areas, where no useful data could be measured, were marked as a checkered area on the plots to keep them distinguishable from analyzable areas (only in the case of the experimental data shown on the right in Figure 5.30 and Figure 5.31).

Figure 5.30 and Figure 5.31 also showed a comparison of the static and total pressure for the two experiments (configurations 1 and 3) and the three PowerFLOW runs (DWT, DIVK configuration 1, and DIVK configuration 3) in the plane "X = 500 mm." The position of the plane, relative to the vehicle, can be seen in both figures on the bottom right. The measurement plane, "X = 500 mm," was placed intentionally to be close to the vehicle base, in order to capture the vehicle's near field wake (approximately at 0 < y < 200 mm, 0 < z < 250 mm). Furthermore, the shear layer generated by turbulence generators at the nozzle exit plane was captured by the plane (600 < y < 1,000 mm).

The plots have been arranged to directly compare the same geometrical situation in simulation and experiment in two rows. The third row only shows one picture which is the blockage-free simulation in the DWT. A comparable measurement for this case was not available. It was displayed here to show differences of the flow around the vehicle in blockage-free and real wind tunnel environment.

For the plots of static pressure (Figure 5.30) and total pressure (Figure 5.31), the following findings were deduced:

The simulations in DIVK both matched the near field wake of the model, as well the shear layer development seen in the measurements. Overall pressure levels seemed to be in good agreement for both cases. The basic flow structure of experimental and DIVK results both matched for configuration with small discrepancies in detail.

The influence of static pressure distribution was clearly visible in Figure 5.30, when comparing the two top rows (configuration 1 versus 3). The static pressure level was higher in configuration 3 (both simulation and experiment), which was due to the static pressure distribution (see Figure 5.22).

Bigger differences were seen in the comparison of DIVK and DWT results. Different pressure levels in the near field wake of the model, as well as the whole surrounding of the model, were visible. In DWT, pressure levels were not influenced by a static pressure distribution and a open-jet shear layer. The static pressure variation shown by DIVK around the car was much closer to the experiment than the DWT static pressure results.

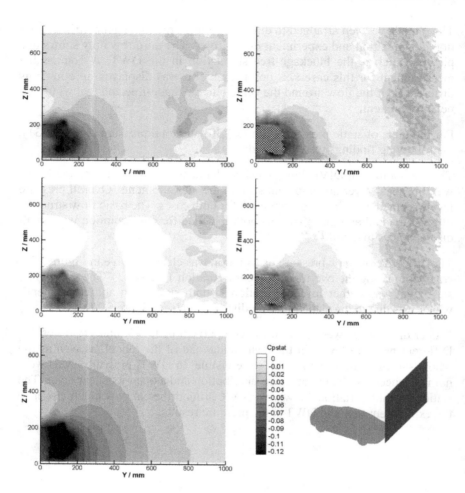

Figure 5.30: Notchback (scale 1:5), $C_{p, stat}$ at X = 500 mm plane: DIVK config 1 (top left); Exp config 1 (top right); DIVK config 3 (center left); Exp config 3 (center right); DWT (bottom left), no measurement data in checkered area

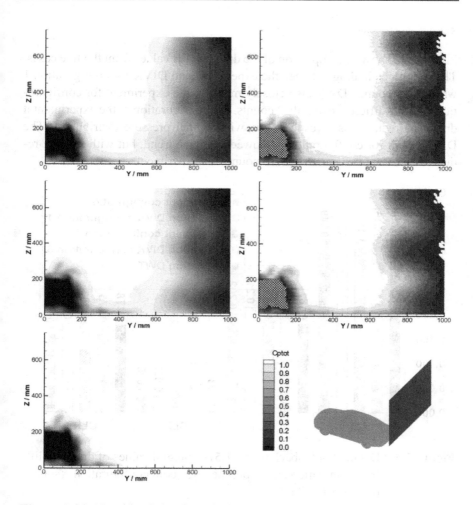

Figure 5.31: Notchback (scale 1:5), $C_{p,\,tot}$ at X = 500 mm plane: DIVK config 1 (top left); Exp config 1 (top right); DIVK config 3 (center left); Exp config 3 (center right); DWT (bottom left), no measurement data in checkered area

Further investigation results of the flow field around the notchback model (scale 1:5) are presented in Appendix A.5.

5.4.2.3   Forces

Figure 5.32 shows a comparison of the drag and lift values from the five cases. The difference in drag between the experiment and DIVK for configuration 1 was only 5 counts. DWT also closely matched the experiment for configuration 1 with a difference of only 2 counts. For configuration 3, the experimental drag was reduced as a result of the wind tunnel pressure distribution. The DIVK result for configuration 3 showed the same trend, but with a larger reduction (26 counts compared to 16 counts for experiment).

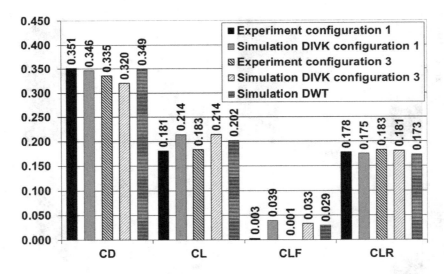

Figure 5.32: Detailed notchback (scale 1:5)—Integral values of drag and lift from simulation and experiment for both configurations (static pressure distribution)

5.4.3   Detailed notchback (scale 1:4)

The notchback vehicle model was a detailed scale model (1:4) with side mirrors and a detailed underbody design, but no underhood flow (closed grille). The blockage ratio in IVK-MWK was 8.75%. Further information, including pictures of the model, can be found in Appendix A.1.3.

The model was tested and simulated in all three configurations shown in Figure 5.22. The DIVK simulation with vehicle model consisted of about 55 million voxels; the DWT simulation used about 50 million voxels. All wind tunnel parts, which were significant for the force measurements, like rocker panel struts and wind tunnel pads, were represented in the computational setup.

5.4.3.1   Flow Field

Figure 5.33 shows the static pressure, respectively, in the Y = 0 mm plane for the different experimental and simulation setups. The simulation data were extracted at the same measurement points as the experimental data to ensure the best comparability. Measurements in the checkered areas were not possible to obtain, due to the limited acceptance cone of the Cobra Probe (see Chapter 5.2).

The plots in Figure 5.33 were arranged to allow for a direct comparison of simulation and experiment for each geometrical situation in each row. The bottom row shows the simulation result of the blockage-free simulation in the DWT. The position of the plane, relative to the vehicle, can be seen on the bottom right.

For the plots in Figure 5.33, the following findings were deduced. The basic flow structure in comparable regions (not checkered areas) was in good agreement between simulations (left) and experiments (right). The overall pressure level was slightly higher in the experiments. The influence of static pressure distribution was clearly visible by comparing plots in different rows. The static pressure level, close to the model, changed with changing wind tunnel pressure distribution; and, this directly influenced the vehicle drag (see Chapter 4.1.2).

As expected, bigger differences were seen with the comparison of DIVK and DWT (bottom left) results. The overall pressure level, and the pressures close to the model base, were lower than in all DIVK simulations. This was due to the missing static pressure distribution in DWT. The basic flow structure in DWT was close to the DIVK results, as was expected. Hence, it was observed that a relatively small change in static pressure distribution would directly affect the pressure level and, thereby, the force measurements; but, it would not drastically change the basic flow structure.

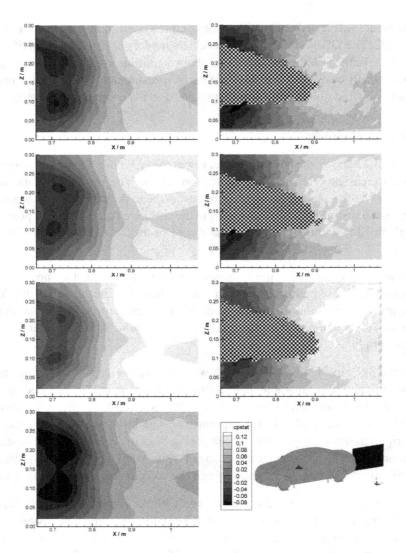

Figure 5.33: Notchback (scale 1:4), Cp-static at Y = 0 mm plane in the near
field wake: top row: DIVK config 1 (left); Exp config 1 (right);
second row: DIVK config 2 (left); Exp config 2 (right); third
row: DIVK config 3 (left); Exp config 3 (right); DWT (bottom
row left), no measurement data in checkered areas

5.4.3.2 Forces

In this section, results from experiment and simulation in the DIVK and the DWT are compared and discussed for the notchback model.

Figure 5.34, Figure 5.35 and Figure 5.36 show a comparison of the drag and lift values for the several cases including the corrected drag value from the experiments.

The deltas between the DWT simulation and uncorrected experimental values, strongly depended on the static pressure distribution, which showed the necessity to account for wind tunnel interference effects.

The DIVK simulations captured the effects of changing static pressure distribution. The difference in drag between the experiment and DIVK was 5 counts for configuration 1. This difference was even smaller for configuration 2 and configuration 3.

In comparision to configuration 1, the experimental drag was reduced for configurations 2 and 3 due to horizontal buoyancy. The DIVK results for configurations 2 and 3 reproduced the drag reduction exactly.

The overall lift was reproduced satisfactorily for all configurations by the simulations. In all cases, the front lift was underestimated by about 12 counts, and the rear lift was overestimated by about 15 counts.

For this model, the changed pressure distribution had a strong effect on drag, but nearly no effect on lift values. This effect was also reproduced for all configurations of the DIVK simulations. It was concluded that the effects of wind tunnel interference on the acting forces, were simulated correctly within the DIVK setup.

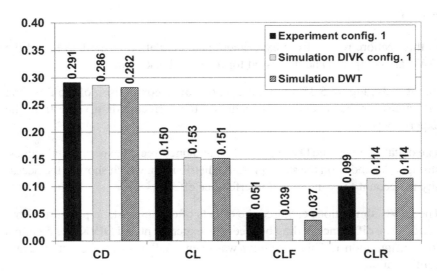

Figure 5.34: Detailed notchback (scale 1:4): Drag and lift values for Experiment, DIVK (configuration 1), and DWT

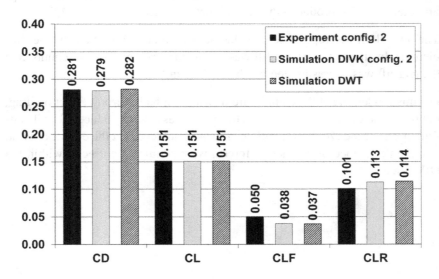

Figure 5.35: Detailed notchback (scale 1:4): Drag and lift values for Experiment, DIVK (configuration 2), and DWT

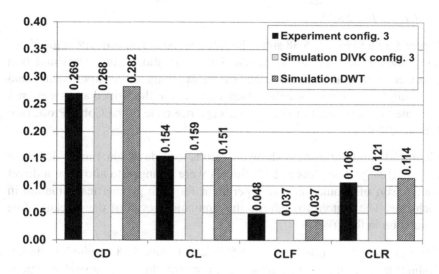

Figure 5.36: Detailed notchback (scale 1:4): Drag and lift values for Experiment, DIVK (configuration 3), and DWT

5.4.4 Detailed SUV model (scale 1:4)

The SUV vehicle model was a detailed scale model (1:4) with side mirrors and a detailed underbody design, but no underhood flow (closed grille). The blockage ratio in IVK-MWK was 10.8%. Further information, including pictures of the model, can be found in Appendix A.1.4.

The SUV model was tested in two different ride heights, which were referred to as "ride height 1" (RH1) and "ride height 2" (RH2). ride height 2 denoted a vehicle configuration with the vehicle body 10 mm lower, as compared to ride height 1. This difference in ride heights equated to 40 mm for the full-scale model. The model was tested and simulated in all three wind tunnel configurations shown in Figure 5.22. The DIVK simulation with the vehicle model consisted of about 67 million voxels; the DWT simulation used about 67 million voxels.

5.4.4.1   Flow Field

Figure 5.37 and Figure 5.38 show the static pressure respectively in the Y = 0 mm plane for the different experimental and simulation setups. The simulation data were extracted at the same measurement points as the experimental data to ensure the best comparability. Measurements in the striped areas were not possible to obtain due to the limited acceptance cone of the Cobra Probe (see Chapter 5.2).

Figure 5.37 and Figure 5.38 show the results for ride height 1 and ride height 2, respectively. The plots in both figures were arranged to allow for a direct comparison of simulation and experiment for each geometrical situation in each row. The bottom row shows the simulation result of the blockage-free simulation in the DWT.

The analysis of the plots in Figure 5.37 and Figure 5.38 resulted in similar findings as in Chapter 6.1.5, as the basic flow structure in comparable regions (not striped areas) was in good agreement between simulations (left) and experiments (right). The overall pressure level fit to the experimental values; and, the influence of static pressure distribution was clearly visible when comparing plots in different rows. The static pressure level, close to the model, changed with changing wind tunnel pressure distribution; and, this directly influenced the vehicle drag (see Chapter 4.1.2).

As already obtained in the above examples, the flow structure in DWT and DIVK results were comparable. The pressures in the DWT simulation results (plots bottom left) correlated with the DIVK (configuration 1) results (plots top left). This finding also correlated with the measured drag forces (see Figure 5.39 and Figure 5.42). For the other two DIVK results (configurations 2 and 3), the overall pressure level and the pressures close to the models base were higher due to the changed static pressure distributions.

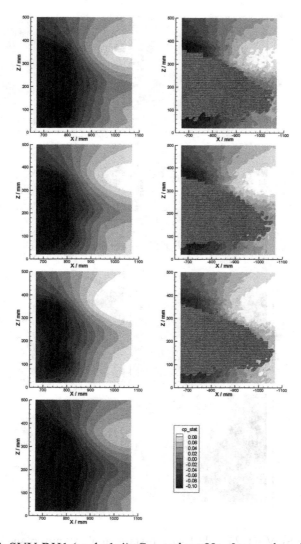

Figure 5.37: SUV-RH1 (scale 1:4), Cp-static at Y = 0 mm plane in the near field wake (no measurement data in striped areas): top row: DIVK config 1 (left); Exp config 1 (right); second row: DIVK config 2 (left); Exp config 2 (right); third row: DIVK config 3 (left); Exp config 3 (right); bottom row: DWT (left)

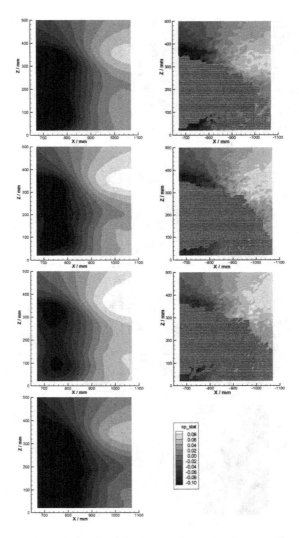

Figure 5.38: SUV-RH2 (scale 1:4), Cp-static at Y = 0 mm plane in the near
field wake (no measurement data in striped areas): top row:
DIVK config 1 (left); Exp config 1 (right); second row: DIVK
config 2 (left); Exp config 2 (right); third row: DIVK config 3
(left); Exp config 3 (right); bottom row: DWT (left)

5.4.4.2 Forces

In this section, results from experiment and simulation in the DIVK and the DWT are compared and discussed for the SUV model, inclusive of the corrected drag values from the experiments. All results are shown in Figure 5.39 to Figure 5.44.

As expected, the experimental drag was reduced for configurations 2 and 3, as compared to configuration 1, as a result of the wind tunnel pressure distribution. This was the case for both ride heights. It must be noted that front lift was negative for all simulation results, while the experimental values were positive. This could be due to an overestimation of the downforce generated by a front spoiler on the model. The underestimated front lift and overestimated rear lift were also observed in the results for the (1:4 scale) notchback (see Chapter 5.4.3).

The deltas between the DWT simulation and the experimental values strongly depend on the wind tunnel configuration and, therefore, the static pressure distribution. This again shows the necessity to take wind tunnel interference effects into account.

The DIVK simulations captured the effects of the changing static pressure distribution for both ride heights. The maximum effect of the changed static pressure distribution was around 30 counts for both ride heights (ride height 1: 28 counts, ride height 2: 34 counts).

For ride height 1, differences in drag between DIVK simulations and experiments were 7 counts for configuration 1, but only 2 counts for the other configurations. For ride height 2, differences in drag between DIVK simulations and experiments were between 2 to 6 counts. Changing the static pressure distribution had no effect on front lift in experiments and simulations. The rear lift was raised between configuration 1 and 3. This effect was captured by the DIVK simulation.

The effects of wind tunnel interference effects on the acting forces were simulated correctly within the DIVK setup.

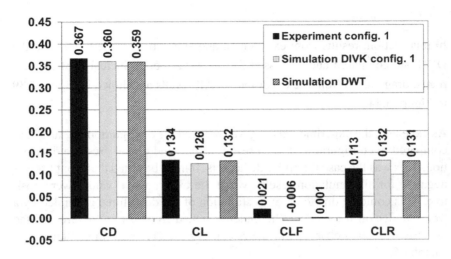

Figure 5.39: Detailed SUV-RH1 (scale 1:4): Drag and lift values for Experiment, DIVK (configuration 1), and DWT

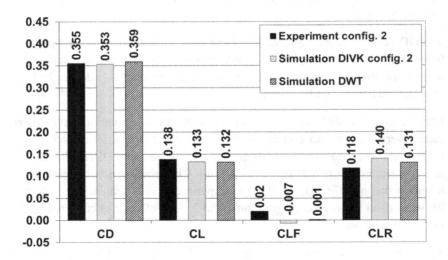

Figure 5.40: Detailed SUV-RH1 (scale 1:4): Drag and lift values for Experiment, DIVK (configuration 2), and DWT

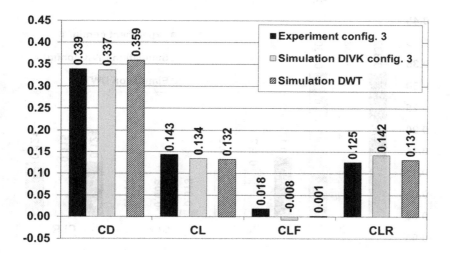

Figure 5.41: Detailed SUV-RH1 (scale 1:4): Drag and lift values for Experiment, DIVK (configuration 3), and DWT

The measured differences between the two ride heights were basically reproduced by the simulation, with the exception of the rear lift. The measured drag reduction, due to lower ride height, was 35-41 counts. The simulated drag reduction due to lower ride height was 30-33 counts. Hence, the drag reduction was underestimated by the simulation by 5-8 counts.

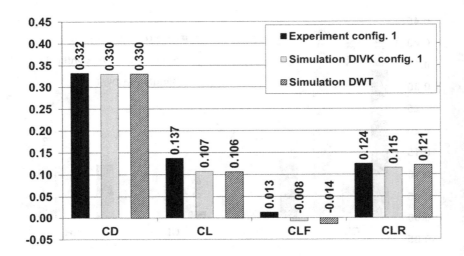

Figure 5.42: Detailed SUV-RH2 (scale 1:4): Drag and lift values for Experiment, DIVK (configuration 1), and DWT

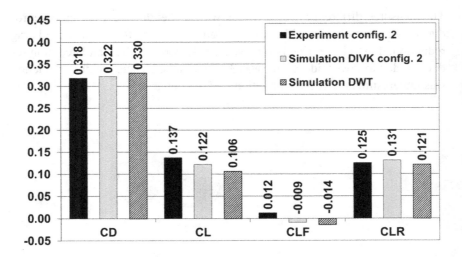

Figure 5.43: Detailed SUV-RH2 (scale 1:4): Drag and lift values for Experiment, DIVK (configuration 2), and DWT

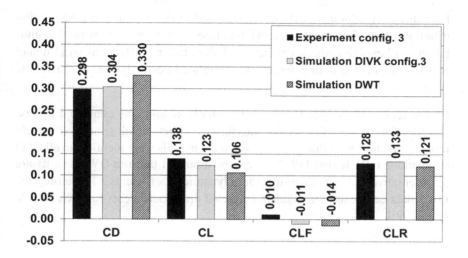

Figure 5.44: Detailed SUV-RH2 (scale 1:4): Drag and lift values for experiment, DIVK (configuration 3), and DWT

5.5 Conclusions

It has been shown by means of a CFD investigation, that horizontal buoyancy can be simulated with PowerFLOW. The correction component for horizontal buoyancy works for static pressure distributions, which can be representative of typical automotive wind tunnels.

For the simulation of the IVK-MWK, the flow conditions in the wind tunnel test section without a vehicle model present were validated by various comparisons between simulations and experiments. The wind tunnel geometry and simulation boundary conditions were setup to match boundary layer, shear layer, and static pressure gradients. This was done to ensure that all interference effects were simulated correctly when a model was placed into the test section. The DIVK was completely validated for further simulations with vehicle models.

Four different vehicle models were simulated in DIVK. Surface pressure distributions, flow field measurements, and forces were validated between simulations and experiments. It can be concluded, that the effects of wind tunnel interference on the acting forces, are simulated correctly within the DIVK setup.

It was demonstrated that the presented DIVK simulations generated a good reproduction of the influences of interference effects on all simulated vehicle models. As the simulation setups close to the models were identical in all CFD simulations (DIVK und DWT), it can be assumed that the DWT results are representative for blockage-free aerodynamic coefficients. Therefore, the DWT data can be used to validate corrected drag coefficients for each vehicle model [60].

6 Application and Investigation of the Correction Method

Within the aerodynamic community today, it is still discussed whether measurements in open-jet wind tunnels need to be corrected due to open-jet interference effects caused by the limited extent of the test section and the jet. Existing correction methods are still not applied in most open-jet wind tunnels to increase comparability. In this section, the correction method, as described in Chapter 4, is applied to measurements from different wind tunnels to show the usability and robustness of the method. Measurements done with full-scale and model-scale vehicles are corrected. Furthermore, a high-blockage situation and an inhomogeneous static pressure distribution are investigated.

6.1 Application to Wind Tunnel Measurements

For the application of the modified Mercker-Wiedemann method, which was been presented above, additional measurements in the wind tunnel were needed, as compared to standard wind tunnel tests. Forces needed to be measured using both plenum and nozzle method. Furthermore, additional measurements were mandated for two different static pressure distributions in the test section.

This section shows several results of the correction method. The measurements were performed in IVK-MWK and the full-scale aero-acoustic wind tunnel at IVK (IVK-FWK). Different methods were used to generate different static pressure distributions. Furthermore, the robustness of the method, concerning high-blockage and inhomogeneous pressure distributions, was investigated.

The correction method was applied to all measurements, both with and without the empirical near field wake constant (see Chapter 4.2.2). This led to two

© Springer Fachmedien Wiesbaden GmbH, part of Springer Nature 2018
O. Fischer, *Investigation of Correction Methods for Interference
Effects in Open-Jet Wind Tunnels*, Wissenschaftliche Reihe
Fahrzeugtechnik Universität Stuttgart, https://doi.org/10.1007/978-3-658-21379-4_6

separate correction results for each data set presented in this chapter. The question, as to which of these results would be considered the right value, will be answered in Chapter 7.

For the SAE Squareback model, used in the first validation of DIVK, measurements in the IVK-MWK with plenum method, were not available (see Chapter 5.4.1). Therefore, the correction method was not be applied to this model.

6.1.1   Coupe Vehicle in IVK-FWK

The coupe vehicle used for this investigation was a standard full-scale production vehicle, and was measured in its standard configuration. The blockage ratio in IVK-FWK was 8.65%. Further information can be found in Appendix A.1.5.

IVK-FWK had a typical static pressure distribution for an automotive open-jet tunnel of this age. The pressure rise in the rear part of the test section was due to the flow stagnation close to the collector. In most wind tunnels, which were built afterward, the test section is longer, relative to the size of nozzle and collector, which moves the pressure rise in front of the collector, further downstream, and away from the measured object. Some of the automotive wind tunnels built recently feature new collector and test section designs, which make it possible to generate static pressure distribution without any gradients in the relevant areas of the test section [48][61][62].

In the past, a variety of methods have been used in various wind tunnels to generate different static pressure distributions [47][63][64]. One example is the stagnation bodies which was used in the model-scale tunnel, as described in Chapter 5.3.5. The advantage of this method was a good homogeneous static pressure distribution. Stagnation bodies were not used by IVK in the full-scale tunnel, due to the huge effort required to apply them in a full-scale tunnel. Moreover, the IVK-FWK collector geometry was as different compared to IVK-MWK. Therefore, it was decided to change the collector flap angle which led to a sufficient change in static pressure to apply the correction method. Figure presents three static pressure distributions in the IVK-FWK with two different collector flap angles and the standard configuration.

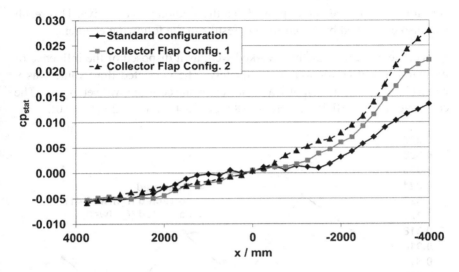

Figure 6.1: Static pressure distributions in IVK-FWK generated by changing the collector flap angle (without ground simulation)

The vehicle was measured in these three different static pressure distributions. Figure 6.2 shows the measured and corrected drag coefficients for the different static pressure distributions and the two different determination methods for dynamic pressure which were described in Chapter 2.1.2. These wind tunnel configurations resulted in a total spread of 14 drag counts over all measured drag values. The influence of using plenum method or nozzle method was evident, as drag values measured with plenum method were always lower, as compared to drag values measured with nozzle method. In this case, the difference was about 6 drag counts. Also shown, was that with rising pressure in front of the collector, drag was reduced, as expected.

The correction method was applied here, both with and without the empirical near field wake constant, which was described in Chapter 4.2.2. In both cases, the correction method reduced the standard deviation over all measured wind tunnel configurations from 5 counts to 0 counts.

The correction method was designed to result in exactly one correction value if only two static pressure distributions were used. In this case, three different static pressure distributions were utilized; therefore, a standard deviation of zero for the corrected values proved that the method was effective. This result was also confirmed by the correction results in the following chapters.

Hence, the correction method worked well, as it disposed of the influence of the wind tunnel configuration completely. The corrected drag value was 9 counts higher when the empirical near field wake constant was set to zero. The corrected drag coefficients were 0.318 (f_{nfw} = 0.41) and 0.327 (f_{nfw} = 0).

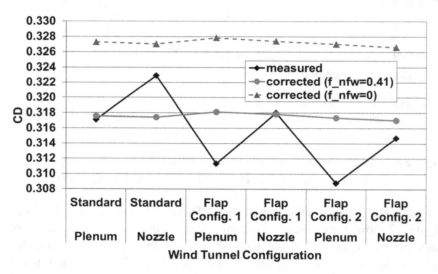

Figure 6.2: Correction results of coupe measured in IVK-FWK (with and without empirical near field wake constant f_{nfw} = 0.41)

6.1.2   Van in IVK-FWK — High-Blockage Setup

The van used for this investigation was a standard full-scale production vehicle, and was measured in its standard configuration (see Figure 6.3). It was chosen because of the big frontal area of nearly 4 m² which resulted in a blockage ratio in IVK-FWK of 17.73%. This would be a fairly high-blockage ratio for standard measurements in most automotive wind tunnels in operation today. Further information on the vehicle can be found in Appendix A.1.6.

Figure 6.3: Van in the test section of IVK-FWK

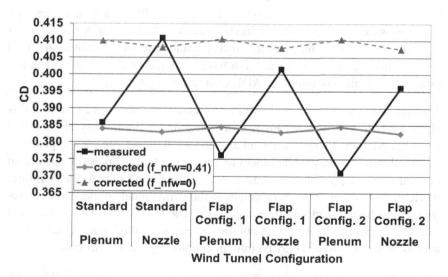

Figure 6.4: Correction results of the van measured in IVK-FWK (with and without the empirical near field wake constant $f_{nfw} = 0.41$)

The vehicle was measured in three different static pressure distributions, as shown in Figure . Figure 6.4 shows the measured and corrected drag coefficients. The results were comparable to the results of the coupe vehicle (see Chapter 6.1.1). The different wind tunnel configurations generated a total spread of 40 drag counts. The difference between using plenum method or nozzle method was about 25 drag counts. The correction method reduced the standard deviation over all wind tunnel configurations from 15 counts to 1 count. It was concluded that the correction method also worked for high-blockage cases. The corrected drag value was about 25 counts higher when the empirical near field wake constant was set to zero. The corrected drag coefficients were 0.384 (f_{nfw} = 0.41) and 0.409 (f_{nfw} = 0).

6.1.3 Detailed Notchback (Scale 1:4) in IVK-MWK — Inhomogeneous Static Pressure Distribution

In order to test the influences of inhomogeneous static pressure distributions in the test section on the correction method, special experiments were conducted in IVK-MWK. A second vehicle model was setup to act as a stagnation body downstream of the test section to generate different inhomogeneous static pressure distributions. The model was placed at several x-positions inside the collector to achieve several different static pressure distributions. This investigation has been published in a white paper for ECARA [65]. The approach was applied within the EADE Correlation Test 2010.

This example for the application of the correction method was chosen for this chapter due to two reasons. First, it demonstrated a simple way of changing the static pressure distribution in an existing wind tunnel. Hence, the possibility of using a second vehicle model as stagnation body makes the correction method easily applicable in almost every existing wind tunnel. Furthermore, this approach required that no changes to the wind tunnel be made and, therefore, could be realized with minimal costs. The second reason was that this example displayed the robustness of the correction method against static pressure distributions, which may be inhomogenous in y- and z-direction.

The vehicle model, which was used to change the static pressure distribution, was the detailed SUV model (scale 1:4 and with closed grille). The ground simulation system of the IVK-MWK was inactive during all measurements.

The wind speed was set to 50 m/s. For the SUV model, the collector blockage ratio amounted to 9.3%. The standard wind tunnel, without any second model present, will be referred to as "Dist_0."

The vehicle model was fixed at four different x-positions behind the turntable inside the collector (see Figure 6.5). The wind tunnel configurations, with the SUV model acting as stagnation body, will be referred to with the running number from "Dist_1" to "Dist_4," while "Dist_1" denotes the smallest change in static pressure distribution and "Dist_4" denotes the biggest change (see Figure 6.6).

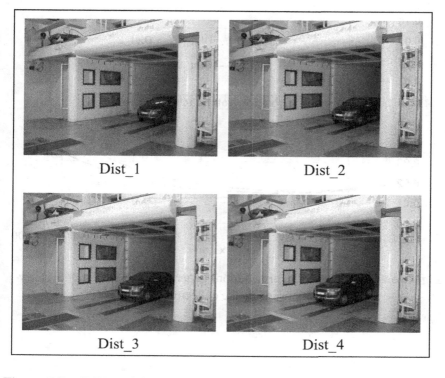

Figure 6.5:   SUV model at several x-positions inside the collector to generate different static pressure distributions in the test section

The static pressure distribution was measured in all spatial directions, parallel to the x-axis, to document the magnitude of static pressure inhomogeneity. The averaged static pressure distributions are displayed in Figure 6.6. Four measurements were taken along the x-axis at the following positions: y = 0 mm and z = 150 mm; y = 300 mm and z = 150 mm; y = 0 mm and z = 450 mm; y = 300 mm and z = 450 mm. The data were averaged over y- and z-direction to generate Figure 6.6.

The inhomogeneity of the static pressure distribution in y-direction is shown in Figure 6.7. This measurement was performed at x = 700 mm, which was directly behind the position where the notchback model was positioned on the underfloor balance.

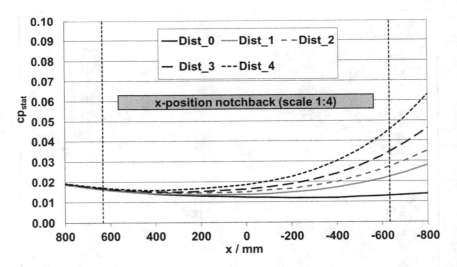

Figure 6.6: Static pressure distributions generated with SUV model at several x-positions inside the collector

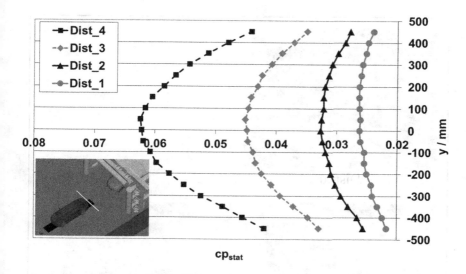

Figure 6.7: Static pressure measurements (x = 700 mm, z = 150 mm); small picture: measurement position (white arrow) relative to vehicles

The measurement setups in the test section for the drag measurements are depicted in Figure 6.8. Figure 6.9 shows the measured and corrected drag coefficients. The results were basically comparable to the results of the coupe (see Chapter 6.1.1) and van (see Chapter 6.1.2). The different wind tunnel configurations generated a total spread of about 55 drag counts. The difference between using plenum method or nozzle method was about 6 drag counts. The correction method reduced the standard deviation over all wind tunnel configurations, from 18 counts to 1 count. This shows that the correction method seemed to also work for measurements performed in inhomogeneous static pressure distributions. The corrected drag value was 9 counts higher when the empirical near field wake constant was set to zero. The corrected drag coefficients were 0.270 (f_{nfw} = 0.41) and 0.279 (f_{nfw} = 0).

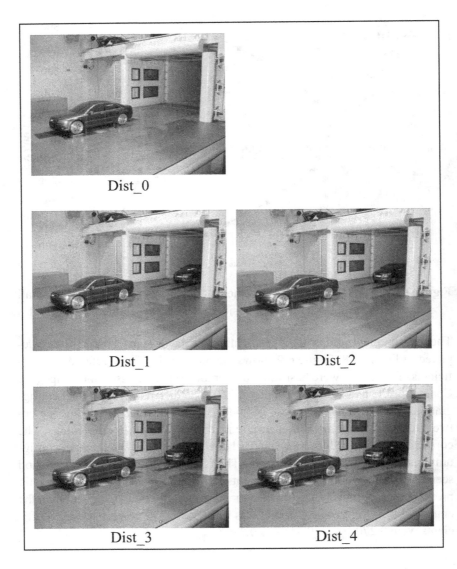

Dist_0

Dist_1   Dist_2

Dist_3   Dist_4

Figure 6.8: Measurement situations in the test section (force measurement
on balance on notchback model; SUV model used as stagnation
body inside collector)

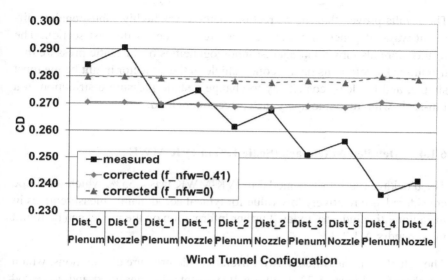

Figure 6.9: Correction results of the notchback model (scale 1:4) in IVK-MWK, in an inhomogenous static pressure distribution (with and without the empirical near field wake constant $f_{nfw} = 0.41$)

In the correction method, the inhomogeneous pressure distribution did not require different treatment from a homogenous distribution. This can be seen in Table 6., as the correction result did not depend on the way the inhomogeneous pressure distribution was used in the correction method. Therefore, one static pressure measurement along the flow direction seemed to be sufficient for each wind tunnel configuration.

Table 6.1: Correction results of notchback model with two different ways of treating the inhomogeneous static pressure distributions in the correction procedure

Static pressure distribution (used for correction method)	CD_{corr}	StDev CD_{corr}
Averaged 4 measurements over y- and z-direction (see Figure 6.6)	0.270	0.001
Single measurement at y = 0 mm, z = 450 mm	0.270	0.001
Delta	0.000	

The results proved that the correction method was highly robust against different ways of generating a static pressure gradient in the test section. The results were also invariant against inhomogeneities of the static pressure distribution. Therefore, using a second vehicle in the collector might be the most simple, and the least costly, way to change a static pressure distribution in a full-scale wind tunnel to apply the correction method.

6.1.4   Detailed Notchback (Scale 1:5) in IVK-MWK

The blockage ratio for this model in IVK-MWK was 5.55%, which would be considered as a relatively low value for typical aerodynamic measurements in the automotive industry today. Further information on the model can be found in Appendix A.1.2.

The vehicle was measured in two different static pressure distributions, which are shown in Figure 5.22. Figure 6.10 presents the measured and corrected drag coefficients. The different wind tunnel configurations generated a total spread of 18 drag counts. The difference between using plenum method or nozzle method was approximately 2 drag counts. The correction method reduced the standard deviation over all wind tunnel configurations from 9 counts to zero.

It must be noted that this was the case here, as the correction method was designed to give exactly one correction value, if only two static pressure distributions were used. But as demonstrated above, the method gave comparable results with more than two static pressure distributions.

The corrected drag value was about 7 counts higher when the empirical near field wake constant was set to zero. The corrected drag coefficients were 0.339 ($f_{nfw} = 0.41$) and 0.346 ($f_{nfw} = 0$).

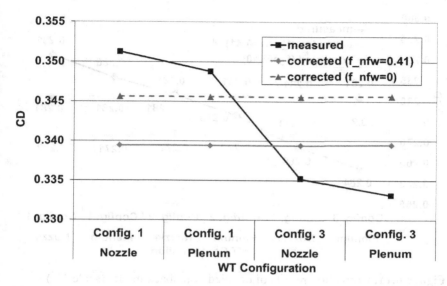

Figure 6.10: Correction results of detailed notchback (scale 1:5) measured in IVK-MWK (with and without the empirical near field wake constant $f_{nfw} = 0.41$)

6.1.5 Detailed Notchback (Scale 1:4) in IVK-MWK

The blockage ratio for this model in IVK-MWK was 8.75%. Further information on the model can be found in Appendix A.1.3. The vehicle was measured in three different static pressure distributions, which are shown in Figure 5.22. Figure 6.11 shows the measured and corrected drag coefficients. The different wind tunnel configurations generated a total spread of 27 drag counts. The difference between using plenum method or nozzle method was about 5 drag counts. The correction method reduced the standard deviation over all wind tunnel configurations from 10 counts to zero. The corrected drag value was about 9 counts higher when the empirical near field wake constant was set to zero. The corrected drag coefficients were 0.272 ($f_{nfw} = 0.41$) and 0.281 ($f_{nfw} = 0$).

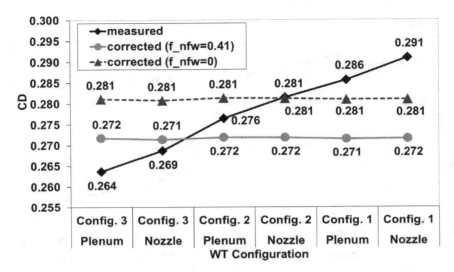

Figure 6.11: Correction results of detailed notchback model (scale 1/4) measured in IVK-MWK (with and without the empirical near field wake constant f_{nfw} = 0.41)

It is essential to state that the same correction result (within 1 drag count) was derived with an inhomogeneous static pressure distribution generated with a second vehicle in the collector. This result was shown in 6.1.3. Therefore, the correction method produced the same correction result, independent from the method used to generate the static pressure distribution.

6.1.6   Detailed SUV Model (Scale 1:4) in IVK-MWK

The blockage ratio in IVK-MWK was 10.8%. As described in Chapter 5.4.4, the SUV model was tested in two different ride heights. The vehicle was measured in three different static pressure distributions, which are shown in Figure 5.22.

Figure 6.12 and Figure 6.13 show the measured and corrected drag coefficients for the two different ride heights. The different wind tunnel configurations generated a total spread of 35 drag counts for ride height 1 and 40 drag counts for ride height 2. The difference between using plenum method or nozzle method was about 7 drag counts, in all cases. The correction method reduced the standard deviation over all wind tunnel configurations from 13 counts to zero for ride height 1. For ride height 2, the standard deviation was reduced from 15 counts to 1 count. The corrected drag value was about 14 counts (ride height 1) and 13 counts (ride height 2) higher when the empirical near field wake constant was set to zero. For ride height 1, the corrected drag coefficients were 0.340 ($f_{nfw} = 0.41$) and 0.354 ($f_{nfw} = 0$). For ride height 2, the corrected drag coefficients were 0.307 ($f_{nfw} = 0.41$) and 0.320 ($f_{nfw} = 0$).

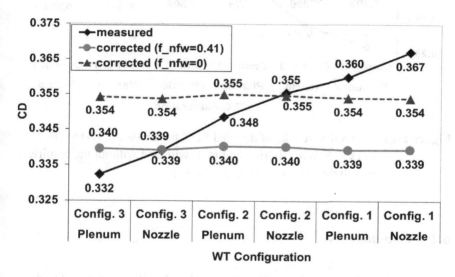

Figure 6.12: Correction results of detailed SUV model (scale 1:4) measured in IVK-MWK—in ride height 1 (with and without the empirical near field wake constant $f_{nfw} = 0.41$)

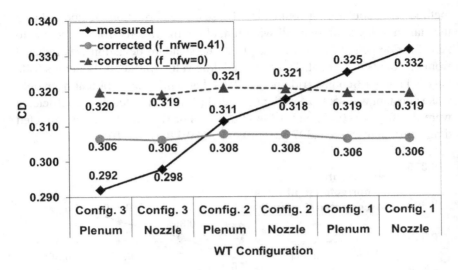

Figure 6.13: Correction results of detailed SUV model (scale 1:4) measured in IVK-MWK—in ride height 2 (with and without the empirical near field wake constant f_{nfw} = 0.41)

6.2 Conclusions

In this chapter, the correction method was applied to measurements of various vehicle shapes and scales in the IVK wind tunnels. Several different ways to generate more than one static pressure distribution in the wind tunnel plenum were used, resulting in relative homogenous and highly inhomogeneous static pressure distributions. The correction method was applied, both with and without the empirical near field wake constant of f_{nfw} = 0.41. Higher values were calculated as the correction results for drag coefficients when the empirical near field wake constant was set to zero (f_{nfw} = 0).

In all cases shown above, the standard deviation over drag coefficients were successfully and drastically reduced to 1 drag count maximum by applying the correction method.

The same correction result was derived for one model (notchback scale 1:4), in two different static pressure distributions (homogenous and inhomogeneous), as shown in Chapter 6.1.5. Therefore, the correction method was very robust against different ways of generating a static pressure gradient; and, the results were also invariant against inhomogeneities of the static pressure distribution.

This research provides evidence and demonstrates that this correction method performed in a highly effective and accurate manner in the version described in this thesis (see details in Appendix A.3).

7 Comparison of Results

In this section, the results of CFD simulations and the corrected experimental results are compared, and the final conclusions of the investigation are derived. In addition, offered are recommendations for future research, which evolved from the performed experiments and simulations.

7.1 Comparison of Computational Fluid Dynamics and Correction Results

In Chapter 5.5, it was concluded that the presented DIVK simulations generated a good reproduction of the influences of interference effects on all simulated vehicle models. Furthermore, it was posited that the DWT simulation results could be representative for blockage-free aerodynamic coefficients for each model [65].

The correction method, discussed in Chapter 6.2, proved to work very well, as it reduced the standard deviation over drag coefficients to 1 drag count maximum. Furthermore, it was learned that correction results were invariant against inhomogeneities of the static pressure distribution in the wind tunnel test.

Therefore, DWT simulation results (see Chapter 5.4) can be directly compared to the corrected drag coefficients (see Chapter 6.1). Comparable data from simulation and correction were available for the following vehicle models: both notchback models (scale 1:5 and 1:4) and the SUV model (scale 1:4, both ride heights). The results of the notchback model (scale 1:4), within an inhomogeneous static pressure distribution, was also comparable.

In Table 7., the data have been compiled for comparison: the vehicle models, the method used to generate static pressure distribution, the blockage-free simulation results of the DWT, and the results from the correction method. In the last two rows of the table, the difference between corrected drag value and the DWT drag value are shown. From these deltas, it can be inferred that corrected drag values (f_{nfw} = 0.41) did not correlate with the DWT results. The deviance

© Springer Fachmedien Wiesbaden GmbH, part of Springer Nature 2018
O. Fischer, *Investigation of Correction Methods for Interference Effects in Open-Jet Wind Tunnels*, Wissenschaftliche Reihe Fahrzeugtechnik Universität Stuttgart, https://doi.org/10.1007/978-3-658-21379-4_7

between corrected drag values and DWT results was strongly reduced by neglecting the empirical near field wake constant ($f_{nfw} = 0$).

Table 7.1:   Comparison of resulting drag values from DWT simulations and the correction method

Model		Notch-back (1:5)	Notch-back (1:4)	SUV ride height 1 (1:4)	SUV ride height 2 (1:4)	Notchback (1:4)
Static pressure distribution		Homogenous (stagnation bodies)				Inhomogeneous (2nd vehicle in collector)
Blockage-free CFD simulations (DWT)		0.349	0.282	0.359	0.330	0.282
Corrected measurement (IVK-MWK)	$f_{nfw} = 0.41$	0.339	0.272	0.340	0.307	0.270
	$f_{nfw} = 0$	0.346	0.281	0.354	0.320	0.279
Delta [corr ($f_{nfw} = 0.41$) – DWT]		-0.010	-0.010	-0.019	-0.023	-0.012
Delta [corr ($f_{nfw} = 0$) – DWT]		-0.003	-0.001	-0.005	-0.010	-0.003

The corrected drag values ($f_{nfw} = 0$) still deviated from the DWT results. For the two notchback models, the devation was reduced to 1-3 drag counts. This small deviation was considered to be a match. For the first SUV case-RH1, the devation was reduced to 5 drag counts, which was a good result. For the second SUV case-RH2, the deviation was reduced to 10 drag counts. A partial explanation for this notable deviation could be that the drag reduction, due to the ride height change, was underestimated in the simulation by 5-8 drag counts (see Chapter 5.4.4). Taking this into account, the drag value from the DWT simulation for SUV-RH2 should have been lower than the value noted in Table 7.. . Therefore, the deviation between corrected drag values ($f_{nfw} = 0$) and DWT results should probably be lower, which would also reduce the deviation.

It was concluded that the correction results ($f_{nfw} = 0$) correlated with the interference-free simulation results of the DWT and, therefore, represent interference-free drag values. Therefore, it was suggested to neglect the term of the correction describing the effect of the near field wake on collector solid blockage ($\Rightarrow f_{nfw} = 0$) in the future.

7.2 Conclusions

The presented data make it possible to compare force coefficients of five different vehicle configurations from experiments and simulations without any influence of wind tunnel interference effects. The comparison of interference-free measures and simulated-force coefficients leads to the conclusion that the near field wake effect in the collector blockage term has a smaller influence than expected in previous publications. Therefore, it is suggested to neglect the term of the correction describing the effect of the near field wake on collector solid blockage ($\Rightarrow f_{nfw} = 0$) in the future. This strongly improves the agreement of interference-free results from experiments and simulations [60] in the IVK model-scale wind tunnel. It has to be investigated in a following thesis, if this can be generalized for all open jet wind tunnels.

This agreement shows that this correction method is able to calculate interference-free drag coefficients. The method works well in the two IVK wind tunnels, which have been used for this thesis. It should be applicable in most aerodynamic wind tunnels, as it was proven to be robust and independent from the method used to generate the second static pressure distribution. This assumption should be investigated in a following thesis.

7.3 Outlook

The work described in this thesis contributes to the ongoing development of a correction method for aerodynamic coefficients measured in open-jet wind tunnels.

The findings presented in this thesis prove that the correction method performs well and can be recommended for widespread and default application in open-jet wind tunnels.

The additional measurements necessary for the correct application of the method should not be underestimated, and will probably prevent a widespread application in the automotive industry. In the meantime, associated projects focus on a simplification of the method, by creating or discovering simple calculation models to make additional measurements unnecessary [66][67].

Further challenges in the field of open-jet correction methods are the application to other open-jet wind tunnels in the automotive industry. This would include the CFD simulation of these wind tunnels to simulate all interference effects. Additional investigations of the flow field could be performed in CFD to help improve the potential-flow model used for the dynamic pressure correction method. Such an approach could also result in a simplification of the simulation of a wind tunnel. This could help reduce the simulation effort, and speed the development pace.

The investigation of different vehicle shapes, other than normal passenger cars as shown above, should be considered for future research projects. An interesting field for investigations would be cars with a high downforce configuration, especially with strong upwash in the near field wake.

8 Bibliography

[1]   Wolf-Heinrich Hucho (Hrsg.), Aerodynamics of Road Vehicles, 4th Edition, Warrendale Pa: SAE, 1998

[2]   Fischer, O., Kuthada, T., Widdecke, N. and Wiedemann, J.: CFD Investigations of Wind Tunnel Interference Effects, SAE Technical Paper 2007-01-1045, Detroit, 2007

[3]   Schröck, D., Widdecke, N. and Wiedemann, J.: On Road Wind Conditions Experienced by a Moving Vehicle. 6th FKFS Conference – Progress in Vehicle Aerodynamics, Stuttgart, 2007

[4]   Mercker, E. and Wiedemann, J.: On the Correction of Interference Effects in Open-jet Wind Tunnels, SAE Technical Paper 960671, Detroit, 1996

[5]   Mercker, E., Wickern, G., Wiedemann, J.: Contemplation of Nozzle Blockage in Open-jet Wind tunnels in View of Different 'Q' Determination Techniques, SAE Technical Paper 970136, Detroit, 1997

[6]   Wickern, G.: On the Application of Classical Wind Tunnel Corrections for Automotive Bodies, SAE Technical Paper 2001-01-0633, Detroit, 2001

[7]   Wickern, G.: Gradient Effects on Drag due to Boundary-Layer Suction in Automotive Wind Tunnels, SAE Technical Paper 2003-01-0655, Detroit, 2003

[8]   Wiedemann, J., Fischer, O., Pang J.: Further Investigations on Gradient Effects, SAE Technical Paper 2004-01-0670, Detroit, 2004

[9]   Wickern, G., Schwartekopp, B.: Correction of Nozzle Gradient Effects in Open-jet Wind Tunnels, SAE Technical Paper 2004-01-0669, Detroit, 2004

[10] Mercker, E., Cooper, K.R., Fischer, O., Wiedemann, J.: The influence of a Horizontal Pressure Distribution on Aerodynamic Drag in Open and Closed Wind Tunnels, SAE Technical Paper 2005-01-0867, Detroit, 2005

© Springer Fachmedien Wiesbaden GmbH, part of Springer Nature 2018
O. Fischer, *Investigation of Correction Methods for Interference
Effects in Open-Jet Wind Tunnels*, Wissenschaftliche Reihe
Fahrzeugtechnik Universität Stuttgart, https://doi.org/10.1007/978-3-658-21379-4

[11] Mercker, E., Cooper, K.R.: A Two-Measurement Correction for the Effects of a Pressure Gradient on Automotive, Open-Jet, Wind Tunnel Measurements, SAE Technical Paper 2006-01-0568, Detroit, 2006

[12] Wetzel, W.: Übertragbarkeit aerodynamischer Beiwerte von Windkanalversuchen im Modellmaßstab auf reale Fahrzeuggeometrien (Validity of aerodynamic coefficients determined from wind tunnel tests in model scale for real full scale vehicle geometries), PhD Thesis, University of Karlsruhe, 2001

[13] Perzon, S.: On Blockage Effects in Wind Tunnels – A CFD Study, SAE Technical Paper 2001-01-0705, Detroit, 2001

[14] Yang, Z., Schenkel, M., Fadler, G.J.: Corrections of the Pressure Gradient Effects on Vehicle Aerodynamic Drag, SAE Technical Paper 2003-01-0935, Detroit, 2003

[15] Yang, Z., Schenkel, M.: Assessment of Closed-Wall Wind Tunnel Blockage Using CFD, SAE Technical Paper 2004-01-0672, Detroit, 2004

[16] Yang, Z., Nastov, A., Schenkel, M.: Further Assessment of Closed-Wall Wind Tunnel Blockage Using CFD, SAE Technical Paper 2005-01-0868, Detroit, 2005

[17] Ewald, B. (ed.): Wind Tunnel Wall Corrections, AGARDograph 336, 1998

[18] Barlow, J.B.,Rae, W.H., Pope, A.: Low-Speed Wind Tunnel Testing, 3rd edition, John Wiley & Sons, inc., 1999

[19] Wuest, W.: Strömungsmeßtechnik, Vieweg-Verlag, 1969

[20] Künstner, R., Deutenbach, K.-R., Vagt, J.-D.: Measurement of Reference Dynamic Pressure in Open-Jet Automotive Wind Tunnels, SAE Technical Paper 920344, Detroit, 1992

[21] Wickern, G.: Recent Literature on Wind Tunnel Test Section Interference Related to Ground Vehicle Testing, SAE Technical Paper 2007-01-1050, Detroit, 2007

[22] Kuthada, T.: CFD and Wind Tunnel – Competitive Tools or Supplementary Use? In: Wiedemann, J., Hucho, W. (Hrsg.): Progress in Vehicle Aerodynamics – Numerical Methods, 5. Short Course at FKFS. Stuttgart, 2006.

[23] Bhatnagar, P., Gross, E. and Krook, M.: A model for collision processes in Gases. I. Small amplitude processes in charged and neutral one-component system, Phys. Rev. Vol. 94, pp. 51-525, 1954.

[24] Chen, H., Teixeira, C. and Molvig, K.: Digital Physics Approach to Computational Fluid Dynamics, Some Basic Theoretical Features, Intl. J. Mod. Phys. C, Vol. 8 (4), pp. 675-684, 1997.

[25] Chen S., Chen, H., Martinez, D. and Matthaeus, W.: Lattice Boltzmann model for simulation of magnetohydrodynamics, Phys. Rev. Lett., Vol. 67 (27), pp. 3776-3779, 1991.

[26] Qian, Y., d'Humieres, D. and Lallemand, P.: Lattice BGK Models for Navier-Stokes Equation, Europhys. Lett., Vol. 17, pp. 479-484, 1992.

[27] Chen, S. and Doolen, G.: Lattice Boltzmann Method for Fluid Flows, Ann. Rev. Fluid Mech., Vol. 30, pp. 329-364, 1998.

[28] Frisch, U., Hasslacher, B. and Pomeau, Y.: Lattice-gas Automata for the Navier-Stokes Equations, Phys. Rev. Lett. Vol. 56 (14), pp. 1505-1508, 1986.

[29] Chen, H., Orszag, S., Staroselsky, I. and Succi, S.: Expanded Analogy between Boltzmann Kinetic Theory of Fluid and Turbulence, J. Fluid Mech., Vol. 519, pp. 301-314, 2004.

[30] Chen, H., Kandasamy, S., Orszag, S., Shock, R., Succi, S. and Yakhot, V.: Extended Boltzmann Kinetic Equation for Turbulent Flows, Science, Vol. 301, pp. 633-636, 2003.

[31] Yakhot, V., Chen, H. A., Staroselsky, I., Qian, Y., Shock, R., Kandasamy, S., Zhang, R., Mallick, S., and Alexander, C.: A New Approach to Modelling Strongly Non-Equilibrium, Time-Dependent Turbulent Flow, Exa internal publication, 2001.

[32] Yakhot, V. and Orszag, S., A.: Renormalization Group Analysis of Turbulence. I. Basic Theory, J. Sci. Comput., Vol. 1 (2), pp. 3-51, 1986.

[33] Yakhot, V., Orszag, S., Thangam, S., Gatski, T. and Speziale, C.: Development of turbulence models for shear flows by a double expansion technique, Phys. Fluids A, Vol. 4 (7), pp. 1510-1520, 1992.

[34] Chen, H.: Extensions in Turbulent Wall Modeling, Exa internal publication, 1998.

[35] Teixeira, C.: Incorporating Turbulence Models into the Lattice-Boltzmann Method, Intl. J. Mod. Phys., Vol. 9 (8), pp. 1159-1175, 1998.

[36] Shock, R., Mallick, S., Chen, H., Yakhot, V., and Zhang, R.: Recent simulation results on 2D NACA airfoils using a lattice Boltzmann based algorithm, AIAA J. Aircraft, Vol. 39 (3), pp. 434-439, 2002.

[37] Cooper, K.R. (ed.): Closed-test-section wind tunnel blockage corrections for road vehicles, SAE SP 1176, Warrendale, 1996

[38] Garner, H.C. (ed.): Subsonic wind tunnel wall corrections, AGARDograph 109, 1966

[39] Lock, C.N.H.: The Interference of a wind tunnel on a symmetrical body, ARC, R&M 1275, 1929

[40] Young, A.D. and Squire, H.B.: Blockage Correction in a closed rectangular tunnel, ARC R&M 1984, 1945

[41] Mercker, E.: A Blockage Correction for Automotive Testing in a Wind Tunnel with Closed Test Section, J. of Wind Eng. And Ind. Aerod., 22, 1986

[42] Aerodynamic Testing of Road Vehicles—Open Throat Wind Tunnel Adjustment, SAE J2071, Detroit, 1990 (Revision 1994 and 1997)

[43] Munk, M.: Some New Aerodynamic Relations. NACA. Report 114, 1921

[44] Glauert, H.: The effect of a Static Pressure Gradient on the Drag of a Body Tested in a Wind Tunnel, ARC, R&M 1158, 1928

[45] Lindener, N. (ed.): Aerodynamic Testing of Road Vehicles in Open-jet Wind Tunnels, SAE SP-1465, Warrendale, 1999

[46] Gürtler, T.: EADE Correlation Test 1999, 18th EADE-Meeting at Mira, Coventry, 2000

[47] Gürtler, T.: Some Remarks on static pressure gradients in wind tunnels with open-jet test sections, 5th ECARA Subgroup Blockage Meeting, Stuttgart, 2001

[48] Wüst, W.: Verdrängungskorrektur für rechteckige Windkanäle bei verschiedenen Strahlbegrenzungen und bei exzentrischer Lage des Modells, Zeitschrift für Flugwissenschaften, Band 9, 1961

[49] Glauert, H.: Wind Tunnel Interference on Wings, Bodies and Airscrews, ARC, R&M 1566, 1933

[50] Maskell, E.: A Theory of Blockage Effects on Bluff Bodies an Stalled Wings in a Closed Wind Tunnel, ARC, R&M 3400, 1961

[51] Duell, E., Muller, S., Yen, J., Ebeling, W., Mercker, E.: Improving Open-jet Wind Tunnel Axial Pressure Gradients. 7th FKFS Conference – Progress in Vehicle Aerodynamics, Stuttgart, 2009

[52] Fischer, O., Mercker, E., Vagt, J., Wiedemann, J.: The Open-jet Correction Method: Status and Application, ECARA Subgroup Blockage Meeting, Stuttgart, 2010

[53] Wiedemann, J. and Potthoff, J.: The New 5-Belt Road Simulation System of the IVK Wind Tunnels, SAE Technical Paper 2003-01-0429, Detroit, 2003

[54] Website: http://www.turbulentflow.com.au/Products/CobraProbe/Co-braProbe.php (valid 08/2017)

[55] Schröck, D.: Implementierung eines Programms zur Durchführung und Auswertung von Stroemungsfeldmessungen im IVK-Modellwindkanal, Diploma Thesis, Institut fuer Verbrennungsmotoren und Kraftfahrwesen, Universität Stuttgart, 2005

[56] Potthoff, J.: Die IVK-Kraftfahrzeug-Windkanalanlage der Universität Stuttgart. Symposium No. T-30-905-056-7 "Aerodynamik des Kraftfahrzeugs", Haus der Technik, Essen, 1987.

[57] Fischer, O., Kuthada, T., Wiedemann, J., Dethioux, P., Mann, R., Duncan, B.: CFD Validation Study for a Sedan Scale Model in an Open-jet Wind Tunnel, SAE Technical Paper 2008-01-0325, Detroit, 2008

[58] Kuthada, T., Schröck, D., Poffhoff, J., Wiedemann, J.: The Effect of Center Belt Roughness on Vehicle Aerodynamics, SAE Technical Paper 2009-01-0776, Detroit, 2009

[59] Schütz, T., Fischer, O., Kuthada, T.: Design and Validation of an SAE-Squareback best practice PowerFLOW setup, internal FKFS report 07/06, 2006

[60] Fischer, O., Kuthada, T., Mercker, E., Wiedemann, J., Duncan, B.: CFD Approach to Evaluate Wind tunnel and Model Setup Effects on Aerodynamic Drag and Lift for Detailed Vehicles, SAE Technical Paper 2010-01-0760, Detroit, 2010

[61] Duell, E., Kharazi, A., Muller, S., Ebeling, W. et al.: The BMW AVZ Wind Tunnel Center, SAE Technical Paper 2010-01-0118, Detroit, 2010

[62] Heidrich, M.: The new Aeroacoustic-Windtunnel at the Mercedes-Benz Technology Center. 9th FKFS Conference – Progress in Vehicle Aerodynamics, Stuttgart, 2013 (Handout, not part of the proceedings)

[63] Kopp, S.: Experimentelle und theorethische Untersuchungen im Hinblick auf eine Blockierungskorrektur für Automobilwindkanäle mit Freistrahl-meßstrecke (Experimental and Theoretical Investigations with Respect to a Blockage Correction for Automotive Wind Tunnels with Open-jet Test Section), Diplomarbeit, TU München, 1997

[64] Herold, J.-H.: Studie eines analytischen Korrekturverfahrens für Freistrahl-Windkanäle (Study of an analytic correction method for open-jet wind tunnels), Studienarbeit, Universität Stuttgart, 2007

[65] Fischer, O., Mercker, E., Vagt, J., Wiedemann, J.: The Open-jet Correction Method: Status and Application, ECARA Subgroup Blockage Meeting, Stuttgart, 2010

[66] Walter, J., Pien, W., Lounsberry, T. and Gleason, M.: On Using Correlations to Eliminate the Second Measurement for Pressure Gradient Corrections. 9th FKFS Conference – Progress in Vehicle Aerodynamics, Stuttgart, 2013

[67] Lounsberry, T. and Walter, J.: Practical Implementation of the Two-Measurement Correction Method in Automotive Wind Tunnels, SAE Technical Paper 2015-01-1530, Detroit, 2015

[68] Gürtler, T.: Ermittlung des Windkanaleinflusses bei der vergleichenden Messung von unterschiedlich skalierten Grundkörpern in verschiedenen Windkanälen. Diplomarbeit IVK, Universität Stuttgart, 1997

A Appendix

A.1 Vehicles and Vehicle Models

A.1.1 SAE Squareback Model

This generic vehicle model is part of a model family known as the SAE reference body or SAE model [58]. With no wheels, it is supported by four cylindrical struts anchored in the rocker panel. The rear end is designed for four interchangeable rear ends (squareback, hatchback, fastback, and notchback), meaning four models are part of the SAE model family. In this thesis, only the squareback model was used.

Table A.1: Technical specifications of SAE squareback model

Vehicle frontal area (A_M)	0.1198	[m²]
Vehicle volume (V_M)	0.1068	[m³]
Vehicle length (l_M)	1.05	[m]
Scale	1:4	[-]

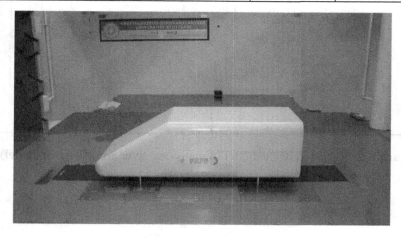

Figure A.1: Picture of SAE squareback model inside IVK-MWK

© Springer Fachmedien Wiesbaden GmbH, part of Springer Nature 2018
O. Fischer, *Investigation of Correction Methods for Interference
Effects in Open-Jet Wind Tunnels*, Wissenschaftliche Reihe
Fahrzeugtechnik Universität Stuttgart, https://doi.org/10.1007/978-3-658-21379-4

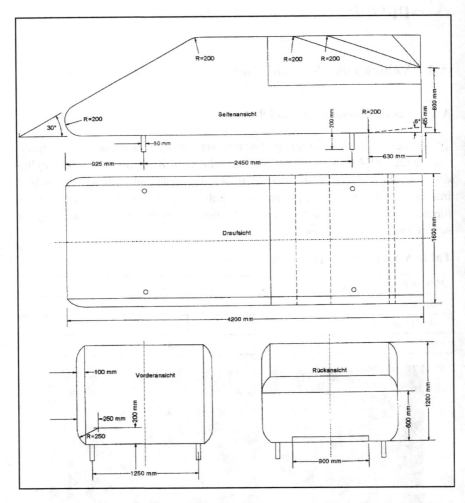

Figure A.2: SAE reference model (dimensions given for full-scale model) [68]

A.1.2   Notchback Model (Scale 1:5)

This notchback model is a detailed CFD validation shell model. It is not equipped for wheel rotation; thus, the model is fixed to the underfloor balance with pins in the tire contact area. It is equipped with pressure taps and does not feature underhood flow (equipped with closed grille).

Table A.2: Technical specifications of notchback model (scale 1:5)

Vehicle frontal area (A_M)	0.09154	[m²]
Vehicle volume (V_M)	0.064236	[m³]
Vehicle length (l_M)	0.98	[m]
Scale	1:5	[-]

Figure A.3: Notchback model (scale 1:5)

A.1.3 Detailed Notchback Model (scale 1:4)

This notchback model is a very detailed presentation shell model. It is equipped for wheel rotation, but no underhood flow (equipped with closed grille). The model is fixed to the balance by rocker panel struts. Forces and moments are measured through both the tire contact area on the wheel rotation unit (WRU) and the rocker panel struts.

Table A.3: Technical specifications of detailed notchback model (scale 1:4)

Vehicle frontal area (A_M)	0.144375	[m²]
Vehicle volume (V_M)	0.119219	[m³]
Vehicle length (l_M)	1.2655	[m]
Scale	1:5	[-]

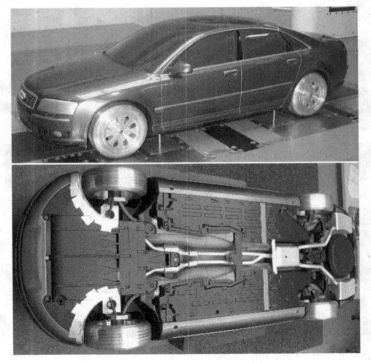

Figure A.4: Detailed notchback model (scale 1:4)

A.1.4   Detailed SUV Model

This SUV model is a very detailed presentation shell model. It is not equipped for wheel rotation and does not feature underhood flow (equipped with closed grille). The model is fixed to the underfloor balance with pins in the tire contact area. For the measurement in IVK-MWK, the WRUs have been removed and replaced by pads. Forces and moments are measured through these pads.

Table A.4:   Technical specifications of detailed SUV model (scale 1:4)

Vehicle frontal area (A_M) — ride height 1	0.179199	[m^2]
Vehicle frontal area (A_M) — ride height 2	0.177923	[m^2]
Vehicle volume (V_M)	0.140625	[m^3]
Vehicle length (l_M)	1.27	[m]
Scale	1:4	[-]

Figure A.5: Detailed SUV model (scale 1:4)

A.1.5   Coupe (full-scale)

The coupe used in these investigations is a Mercedes-Benz (MB) CLK (W208). It is a standard production vehicle.

Table A.5:   Technical specifications of MB CLK

Vehicle frontal area (A_M)	1.942	[m^2]
Vehicle volume (V_M)	6.723	[m^3]
Vehicle length (l_M)	4.567	[m]
Scale	1:1	[-]

A.1.6   Van (full-scale)

The van used in these investigations is a standard production vehicle (Renault Master, Model I).

Figure A.6: Renault Master I

Table A.6:   Technical specifications of Renault Master

Vehicle frontal area (A_M)	3.982	[m²]
Vehicle volume (V_M)	25.213	[m³]
Vehicle length (l_M)	4.89	[m]
Scale	1:1	[-]

A.2   Wind tunnels

A.2.1   IVK Model-Scale Wind Tunnel (IVK-MWK)

The IVK-MWK is an open-jet, Göttingen-type wind tunnel, with a single return duct. It is equipped with a five-belt ground simulation system [56][58].

Table A.7:   Technical specifications of IVK-MWK

Nozzle cross-section (A_N)	1.65	[m²]
Collector cross-section (A_C)	1.949	[m²]
Test section length (l_{TS})	2.584	[m]
Tunnel Factor (τ)	-0.275	[-]
Distance nozzle exit plane – wheelbase center (x_M)	1.225	[m]
Maximum velocity	80	[m/s]

A constructional drawing of the wind tunnel and pictures from the test section are shown in Figure A.7, Figure A.8, and Figure A.9.

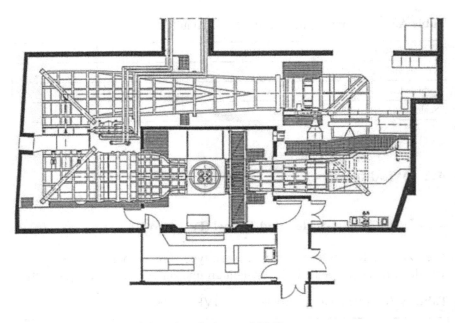

Figure A.7: IVK model-scale wind tunnel [56]

Figure A.8: Nozzle of IVK model-scale wind tunnel

Figure A.9: Collector of IVK model-scale wind tunnel

The definition of the coordinate system in IVK-MWK is shown in Figure A.**10**. The origin (0/0/0) is located in the center of the turntable.

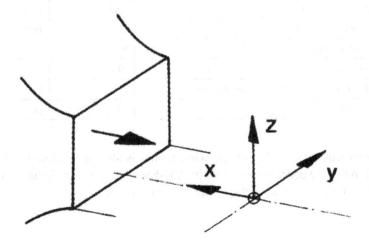

Figure A.10: Coordinate system in IVK-MWK

In this thesis, three different wind tunnel configurations were used throughout all measurements. Stagnation bodies and taped breathers were used to generate these three configurations. The exact positions can be found in Table A.**8**.

Table A.8: Detailed information on stagnation body positions inside the collector

Wind tunnel configuration	Stagnation bodies		Big breathers
	Distance (x-direction) to collector plane	Height	
1	-	-	taped
2	1125	64	taped
3	1125	76	taped

A.2.2   IVK Full-Scale Aero-Acoustic Wind Tunnel (IVK-FWK)

The IVK-FWK is an open-jet Göttingen-type wind tunnel with a single return duct. It is equipped with a five-belt ground simulation system [56][58].

Table A.9: Technical specifications of IVK-FWK

Nozzle cross-section (A_N)	22.5	[m^2]
Collector cross-section (A_C)	26.5	[m^2]
Test section length (l_{TS})	9.9	[m]
Tunnel Factor (τ)	-0.275	[-]
Distance nozzle exit plane – wheelbase center (x_M)	4.5	[m]
Maximum velocity	71	[m/s]

A constructional drawing of the wind tunnel is shown in Figure A.**11**. The definition of the coordinate system in IVK-FWK is shown in Figure A.**12**. The origin (0/0/0) is located in the center of the turntable.

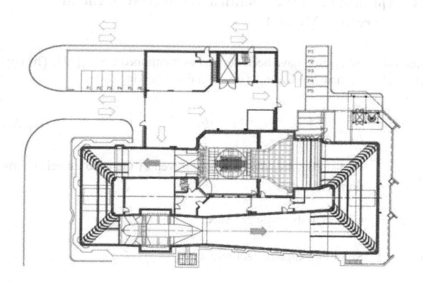

Figure A.11:IVK-FWK [56]

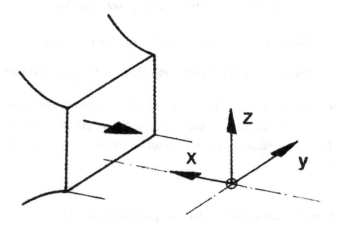

Figure A.12: Coordinate system in IVK-FWK

A.3 Application of the Modified Mercker-Wiedemann Correction Method

Three values of epsilon have been defined as mentioned above [11]. The correction of the dynamic pressure is given in the following form:

$$\frac{q_{corr.}}{q_\infty} = \left[1 + \varepsilon_S + \varepsilon_C + \varepsilon_N + \left(\varepsilon_P - \varepsilon_N\right)\cdot f_P\right]^2 \qquad \text{Eq. A.1}$$

q_∞ : Undisturbed dynamic pressure (given by the wind tunnel calibration)

$q_{corr.}$: Corrected dynamic pressure

ε_S : Solid-blockage correction factor

ε_C : Collector effect correction factor

ε_N : Nozzle effect correction factor (nozzle method)

ε_P : Nozzle effect correction factor (plenum method)

f_P : Plenum factor: Plenum Method: $f_P = 1$, Nozzle Method: $f_P = 0$

The application of the method is separated into two sequential steps described in the following paragraphs. In the first step, the singularity location x_S is calculated [10]. In the second step, the effect of the static pressure distribution is corrected.

A.3.1 Step 1: Calculation of source point position xS

One x_S exists for each measurement done with plenum method and nozzle method.

Table A.10: Correction method: Overview of calculated x_S

Static pressure distribution	q Determination method	Calculated x_S
$cp_{stat\_1}(x)$	nozzle	$x_{S,1}$
$cp_{stat\_1}(x)$	plenum	
$cp_{stat\_2}(x)$	nozzle	$x_{S,2}$
$cp_{stat\_2}(x)$	plenum	

Equivalent nozzle radius: $r_N = \sqrt{\dfrac{2A_N}{\pi}}$ Eq. A.2

Equivalent collector radius: $r_C = \sqrt{\dfrac{2A_C}{\pi}}$ Eq. A.3

$$\varepsilon_{QN,1/2} = \frac{\dfrac{A_M}{2A_N}\left(1 - \dfrac{x_{S,1/2}}{\sqrt{x_{S,1/2}^2 + r_N^2}}\right)}{1 - \dfrac{A_M}{2A_N}\left(1 - \dfrac{x_{S,1/2}}{\sqrt{x_{S,1/2}^2 + r_N^2}}\right)}$$ Eq. A.4

$$\varepsilon_{QP,1/2} = \frac{\dfrac{A_M}{2\pi}\left(\dfrac{x_{S,1/2}}{\sqrt{x_{S1,2}^2 + r_N^2}^{\,3}}\right)}{1 - \dfrac{A_M}{2\pi}\left(\dfrac{x_{S,1/2}}{\sqrt{x_{S,1/2}^2 + r_N^2}^{\,3}}\right)}$$ Eq. A.5

Nozzle blockage:

$$\varepsilon_{N,1/2} = \frac{\varepsilon_{QN,1/2} \cdot r_N^3}{\left(r_N^2 + \left(x_M - \frac{l_M}{2}\right)^2\right)^{3/2}}$$

Eq. A.6

$$\varepsilon_{P,1/2} = \frac{\varepsilon_{QP,1/2} \cdot r_N^3}{\left(r_N^2 + \left(x_M - \frac{l_M}{2}\right)^2\right)^{3/2}}$$

Eq. A.7

Iterative calculation—Variation of $x_{S,1/2}$ until the formulas Eq. A.8 are valid:

$$\frac{(1+\varepsilon_{P,1})^2}{CD_{meas1,P}} = \frac{(1+\varepsilon_{N,1})^2}{CD_{meas1,N}} \quad \frac{(1+\varepsilon_{P,2})^2}{CD_{meas2,P}} = \frac{(1+\varepsilon_{N,2})^2}{CD_{meas2,N}}$$

Eq. A.8

$$CD_{meas,P} = \frac{F_P}{A_M \cdot q_P} \quad CD_{meas,N} = \frac{F_N}{A_M \cdot q_N}$$

Eq. A.9

If forces for nozzle and plenum method are not measured simultaneously (which means $F_N \neq F_P$), the following formulas should be used:

$$(1+\varepsilon_{P,1})^2 \cdot q_{1,P} = (1+\varepsilon_{N,1})^2 \cdot q_{1,N}$$
$$(1+\varepsilon_{P,2})^2 \cdot q_{2,P} = (1+\varepsilon_{N,2})^2 \cdot q_{2,N}$$

Eq. A.10

A.3.2   Step 2: Calculation of Sensitivity Length

It is assumed in the equations that the model

$$\text{X-Position model front: } x_{MF} = x_M - \frac{l_M}{2} \qquad\qquad \text{Eq. A.11}$$

$$\text{X-Position wake end: } x_{WE} = x_M + \frac{l_M}{2} + l_S \qquad\qquad \text{Eq. A.12}$$

$$\Delta CD_{HB1} = cp_{stat1}(x_{WE}) - cp_{stat1}(x_{MF}) \qquad\qquad \text{Eq. A.13}$$

$$\Delta CD_{HB2} = cp_{stat2}(x_{WE}) - cp_{stat2}(x_{MF}) \qquad\qquad \text{Eq. A.14}$$

Wake blockage:

$$\varepsilon_{W,N,1} = \frac{A_M}{A_C}\left(\frac{CD_{meas1,N}}{4} + f_{nfw}\right)$$

$$\varepsilon_{W,P,1} = \frac{A_M}{A_C}\left(\frac{CD_{meas1,P}}{4} + f_{nfw}\right)$$

$$\varepsilon_{W,N,2} = \frac{A_M}{A_C}\left(\frac{CD_{meas2,N}}{4} + f_{nfw}\right) \qquad\qquad \text{Eq. A.15}$$

$$\varepsilon_{W,P,2} = \frac{A_M}{A_C}\left(\frac{CD_{meas2,P}}{4} + f_{nfw}\right)$$

$$f_{nfw} = \begin{cases} 0,41 & \text{near field wake constant according to [4]} \\ 0 & \text{suggestion in chapter 7.2} \end{cases}$$

Collector blockage:

$$\varepsilon_{C,N/P,1/2} = \frac{\varepsilon_{W,N/P,1/2} \cdot r_C^3}{\left(r_C^2 + \left(l_{TS} - x_M - \dfrac{l_M}{2}\right)^2\right)^{3/2}}$$

Eq. A.16

Jet expansion:

$$\varepsilon_S = \frac{\tau \cdot A_M \sqrt{\dfrac{V_M}{l_M}}}{\left(\dfrac{A_N}{1 + \varepsilon_{QN,1/2}}\right)^{3/2}}$$

Eq. A.17

Correction of the dynamic pressure q:

$$\frac{q_{corr.,N1}}{q_\infty} = \left[1 + \varepsilon_{S,1} + \varepsilon_{C,N,1} + \varepsilon_{N,1}\right]^2$$

$$\frac{q_{corr.,P1}}{q_\infty} = \left[1 + \varepsilon_{S,1} + \varepsilon_{C,P,1} + \varepsilon_{P,1}\right]^2$$

Eq. A.18

$$\frac{q_{corr.,N2}}{q_\infty} = \left[1 + \varepsilon_{S,2} + \varepsilon_{C,N,2} + \varepsilon_{N,2}\right]^2$$

$$\frac{q_{corr.,P2}}{q_\infty} = \left[1 + \varepsilon_{S,2} + \varepsilon_{C,P,2} + \varepsilon_{P,2}\right]^2$$

Correction of the drag CD:

$$CD_{corr,N1} = \frac{CD_{meas1,N} + \Delta CD_{HB1}}{\dfrac{q_{corr.,N1}}{q_\infty}}$$

$$CD_{corr,P1} = \frac{CD_{meas1,P} + \Delta CD_{HB1}}{\dfrac{q_{corr.,P1}}{q_\infty}}$$

$$CD_{corr,N2} = \frac{CD_{meas2,N} + \Delta CD_{HB2}}{\dfrac{q_{corr.,N2}}{q_\infty}}$$

Eq. A.19

$$CD_{corr,P2} = \frac{CD_{meas2,P} + \Delta CD_{HB2}}{\dfrac{q_{corr.,P2}}{q_\infty}}$$

Iterative calculation by variation of x_S until the standard deviation over $CD_{corr,N/P,1/2}$ is minimal and the following formula is valid:

$$CD_{corr} = CD_{corr,N1} = CD_{corr,P1} = CD_{corr,N2} = CD_{corr,P2} \qquad \text{Eq. A.20}$$

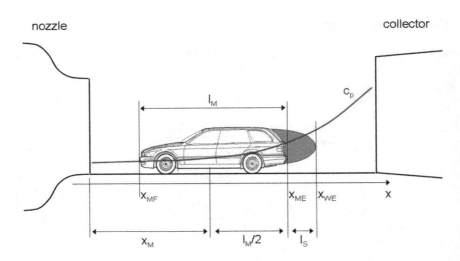

Figure A.13: Illustration of the wind tunnel test section with a model present

A_N	:	Nozzle cross section	[m²]
A_C	:	Collector cross section	[m²]
l_{TS}	:	Jet length	[m]
τ	:	Tunnel factor	[-]
x_M	:	x-position model wheelbase center (distance of wheelbase center to nozzle exit plane)	[m]
l_M	:	Model length	[m]
V_M	:	Model volume	[m³]
A_M	:	Model area frontal	[m²]
f_P	:	Plenum Factor (plenum method: $f_p = 1$, nozzle method: $f_p = 0$)	[-]

$cp_{stat1}(x)$	:	Static pressure distribution 1 in test section	[-]
$cp_{stat2}(x)$	:	Static pressure distribution 2 in test section	[-]
$CD_{meas1,N}$	:	Measured drag value in cpstat_1(x) with nozzle method	[-]
$CD_{meas1,P}$	:	Measured drag value in cpstat_1(x) with plenum method	[-]
$CD_{meas2,N}$	:	Measured drag value in cpstat_2(x) with nozzle method	[-]
$CD_{meas2,P}$	:	Measured drag value in cpstat_2(x) with plenum method	[-]
x_S	:	Source point position	[m]
l_S	:	Sensitivity length	[m]
x_{MF}	:	x-position model front	[m]
x_{ME}	:	x-position model end	[m]
x_{WE}	:	x-position model wake end	[m]

A.4   Calculation of Mass Flow for Boundary Conditions in CFD

The mass flow has been calculated using the ideal gas law:

$$pV = nRT \quad n = \frac{m}{M} \qquad\qquad \text{Eq. A.21}$$

$$\delta = \frac{m}{V} \qquad\qquad \text{Eq. A.22}$$

$$\delta = \frac{p \cdot nM}{nRT} = \frac{pM}{RT} \qquad\qquad \text{Eq. A.23}$$

$$\dot{m} = \delta \cdot v \cdot A_N \qquad\qquad \text{Eq. A.24}$$

$$M = 28.97 \frac{\text{kg}}{\text{kmol}} \quad T = \left(20 + 273.15\right)\text{K} \qquad\qquad \text{Eq. A.25}$$

$$R = 8.3144 \frac{\text{J}}{\text{mol} \cdot \text{K}} \quad p = 101300\,\text{Pa} \qquad\qquad \text{Eq. A.26}$$

p :   Static pressure

V :   Volume

n :   Number of moles

R :   Universal gas constant

T :   Temperature

m :   Mass

M :   Molecular weight

δ :   Density

A_N :   Nozzle cross-section

v :   Wind velocity

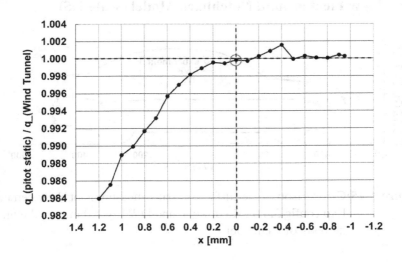

Figure A.14: Ratio of dynamic pressures q in the empty wind tunnel test section

IVK-MWK shows a slight acceleration of the flow coming out of the nozzle. The dynamic pressure has been measured with a pitot static at several x-positions in the symmetry plane of the test section. The ratio of this measured dynamic pressure, and the dynamic pressure given by the wind tunnel, is plotted in Figure A.**14**. As shown, the calibration of the wind tunnel works well, as the ratio is 1 at the calibration position (x = 0).

The calculation of the mass flow shown above gives a mass flow which generates the chosen velocity v at the nozzle exit plane. Due to the slight acceleration of the flow up to the calibration position, the wind velocity v for the mass flow calculation had to be set slightly lower than the chosen wind speed of 50 m/s. The calculated mass flow has been calculated to:

$$\dot{m} = 97.9077 \frac{\text{kg}}{\text{s}}$$

Eq. A.27

A.5 Flow Field around Notchback Model (scale 1:5)

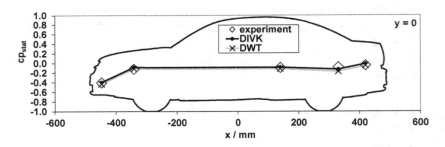

Figure A.15: Comparison of the surface pressure measurements between the Exp config 3, DIVK config 3, and DWT, y = 0 mm on the underbody of the model

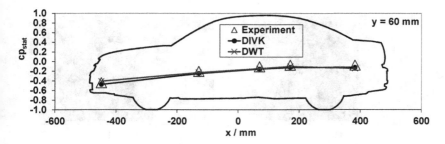

Figure A.16: Comparison of the surface pressure measurements between the Exp config 3, DIVK config 3, and DWT, y = 60 mm on the underbody of the model

The figures in this section show comparisons of further flow field measurements with simulation data for the notchback model [57]. The simulation data were extracted at exactly the same measurement points as the experimental data.

Figure A.17 shows the static pressure, respectively, in the Y = 255 mm plane for the different experimental and simulation setups[57]. The low and high static pressure regions are captured by both the DWT and DIVK. The size of the low pressure regions in the DIVK is more comparable to the experimental results than the DWT results. The influence of the static pressure distribution is visible in the configuration 3 cases. Furthermore, the boundary layer development in the three cases can be clearly seen in Figure 7, where both DIVK and DWT simulations show comparable results.

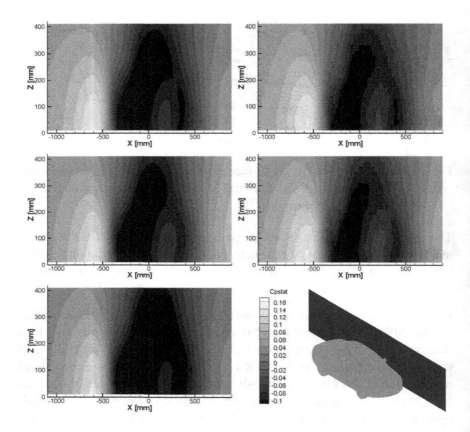

Figure A.17: Cp-static at Y = 255 mm plane: top row: DIVK config 1 (left);
Exp config 1 (right); center row: DIVK config 3 (left); Exp
config 3 (right); bottom row: DWT (left)

Figure A.18 shows the static pressure in the Z = 30 mm plane for the five set-ups [57]. Both DWT and DIVK show the low and high pressure regions similar to the experiment. Again, the influence of the static pressure distribution is visible in the configuration 3 setups.

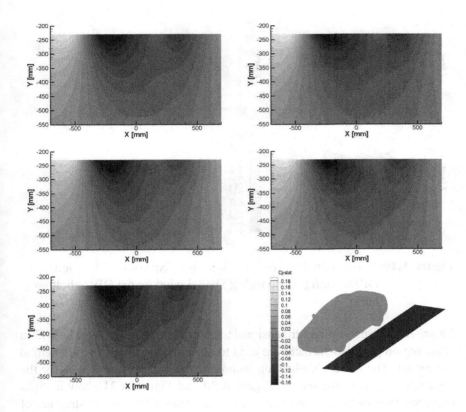

Figure A.18: Cp-static at Z = 30 mm plane: top row: DIVK config 1 (left); Exp config 1 (right); center row: DIVK config 3 (left); Exp config 3 (right); bottom row: DWT (left)

Figure A.19 shows the static pressure distribution in the near field wake at Y = 0 plane for configuration 3 [57]. The distribution predicted by DIVK and DWT shows the influence of the wind tunnel pressure distribution on vehicle drag. The static pressure behind the vehicle is lower in DWT, leading to the increased drag, as compared to configuration 3. For DIVK, the increase in static pressure behind the vehicle, as compared to the experiment, is consistent with the lower drag (see Figure 5.32).

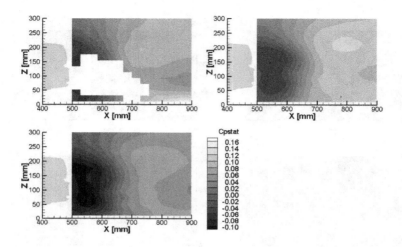

Figure A.19: Cp-static in the wake (Y = 0 plane): Exp config 3 (top left);
DIVK config 3 (top right); Digital wind tunnel DWT (bottom
left)

Figure A.20 displays three-dimensional total pressure plots from the near field
wake region [57]. Iso-surfaces are used to visualize the flow structure behind
the model. The same visualization technique is used in Figure A.21 for the
backlight region of the model. Figure A.20 and Figure A.21 show a slight
overprediction of the a-pillar vortex in the simulation. The overall structure of
both near field wake and backlight area is captured by both DIVK and DWT
simulations.

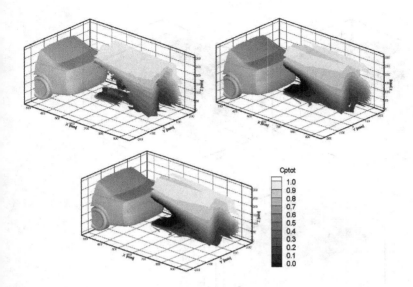

Figure A.20: Cp-total in the wake (positive y): Exp config 3 (top left); Wind tunnel todel DIVK config 3 (top right); Digital wind tunnel DWT (bottom left)

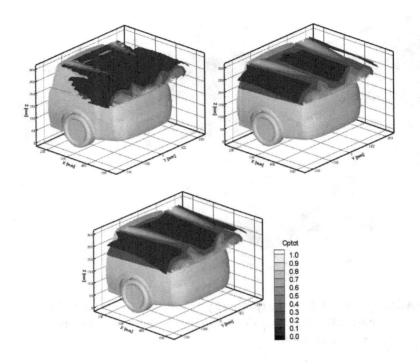

Figure A.21: Cp-total in the backlight area (iso-surfaces): Exp config 3 (top left); Wind tunnel todel DIVK config 3 (top right); Digital wind tunnel DWT (bottom left)

Printed in the United States
By Bookmasters

Printed in the United States
By Bookmasters